INTERNATIONAL SERIES OF MONOGRAPHS ON PHYSICS

SERIES EDITORS

J. BIRMAN	CITY UNIVERSITY OF NEW YORK
S. F. EDWARDS	UNIVERSITY OF CAMBRIDGE
R. FRIEND	UNIVERSITY OF CAMBRIDGE
M. REES	UNIVERSITY OF CAMBRIDGE
D. SHERRINGTON	UNIVERSITY OF OXFORD
G. VENEZIANO	CERN, GENEVA

INTERNATIONAL SERIES OF MONOGRAPHS ON PHYSICS

144. T. R. Field: *Electromagnetic scattering from random media*
143. W. Götze: *Complex dynamics of glass-forming liquids – a mode-coupling theory*
142. V.M. Agranovich: *Excitations in organic solids*
141. W.T. Grandy: *Entropy and the time evolution of macroscopic systems*
140. M. Alcubierre: *Introduction to 3 + 1 numerical relativity*
139. A. L. Ivanov, S. G. Tikhodeev: *Problems of condensed matter physics – quantum coherence phenomena in electron-hole and coupled matter-light systems*
138. I. M. Vardavas, F. W. Taylor: *Radiation and climate*
137. A. F. Borghesani: *Ions and electrons in liquid helium*
136. C. Kiefer: *Quantum gravity, Second edition*
135. V. Fortov, I. Iakubov, A. Khrapak: *Physics of strongly coupled plasma*
134. G. Fredrickson: *The equilibrium theory of inhomogeneous polymers*
133. H. Suhl: *Relaxation processes in micromagnetics*
132. J. Terning: *Modern supersymmetry*
131. M. Mariño: *Chern-Simons theory, matrix models, and topological strings*
130. V. Gantmakher: *Electrons and disorder in solids*
129. W. Barford: *Electronic and optical properties of conjugated polymers*
128. R. E. Raab, O. L. de Lange: *Multipole theory in electromagnetism*
127. A. Larkin, A. Varlamov: *Theory of fluctuations in superconductors*
126. P. Goldbart, N. Goldenfeld, D. Sherrington: *Stealing the gold*
125. S. Atzeni, J. Meyer-ter-Vehn: *The physics of inertial fusion*
123. T. Fujimoto: *Plasma spectroscopy*
122. K. Fujikawa, H. Suzuki: *Path integrals and quantum anomalies*
121. T. Giamarchi: *Quantum physics in one dimension*
120. M. Warner, E. Terentjev: *Liquid crystal elastomers*
119. L. Jacak, P. Sitko, K. Wieczorek, A. Wojs: *Quantum Hall systems*
118. J. Wesson: *Tokamaks, Third edition*
117. G. Volovik: *The Universe in a helium droplet*
116. L. Pitaevskii, S. Stringari: *Bose–Einstein condensation*
115. G. Dissertori, I.G. Knowles, M. Schmelling: *Quantum chromodynamics*
114. B. DeWitt: *The global approach to quantum field theory*
113. J. Zinn-Justin: *Quantum field theory and critical phenomena, Fourth edition*
112. R.M. Mazo: *Brownian motion – fluctuations, dynamics, and applications*
111. H. Nishimori: *Statistical physics of spin glasses and information processing – an introduction*
110. N.B. Kopnin: *Theory of nonequilibrium superconductivity*
109. A. Aharoni: *Introduction to the theory of ferromagnetism, Second edition*
108. R. Dobbs: *Helium three*
107. R. Wigmans: *Calorimetry*
106. J. Kbler: *Theory of itinerant electron magnetism*
105. Y. Kuramoto, Y. Kitaoka: *Dynamics of heavy electrons*
104. D. Bardin, G. Passarino: *The Standard Model in the making*
103. G.C. Branco, L. Lavoura, J.P. Silva: *CP Violation*
102. T.C. Choy: *Effective medium theory*
101. H. Araki: *Mathematical theory of quantum fields*
100. L. M. Pismen: *Vortices in nonlinear fields*
99. L. Mestel: *Stellar magnetism*
98. K. H. Bennemann: *Nonlinear optics in metals*
96. M. Brambilla: *Kinetic theory of plasma waves*
94. S. Chikazumi: *Physics of ferromagnetism*
91. R. A. Bertlmann: *Anomalies in quantum field theory*
90. P. K. Gosh: *Ion traps*
88. S. L. Adler: *Quaternionic quantum mechanics and quantum fields*
87. P. S. Joshi: *Global aspects in gravitation and cosmology*
86. E. R. Pike, S. Sarkar: *The quantum theory of radiation*
83. P. G. de Gennes, J. Prost: *The physics of liquid crystals*
73. M. Doi, S. F. Edwards: *The theory of polymer dynamics*
69. S. Chandrasekhar: *The mathematical theory of black holes*
51. C. Møller: *The theory of relativity*
46. H. E. Stanley: *Introduction to phase transitions and critical phenomena*
32. A. Abragam: *Principles of nuclear magnetism*
27. P. A. M. Dirac: *Principles of quantum mechanics*
23. R. E. Peierls: *Quantum theory of solids*

Electromagnetic Scattering from Random Media

Timothy R. Field
McMaster University

OXFORD
UNIVERSITY PRESS

OXFORD
UNIVERSITY PRESS

Great Clarendon Street, Oxford OX2 6DP

Oxford University Press is a department of the University of Oxford.
It furthers the University's objective of excellence in research, scholarship,
and education by publishing worldwide in

Oxford New York

Auckland Cape Town Dar es Salaam Hong Kong Karachi
Kuala Lumpur Madrid Melbourne Mexico City Nairobi
New Delhi Shanghai Taipei Toronto

With offices in

Argentina Austria Brazil Chile Czech Republic France Greece
Guatemala Hungary Italy Japan Poland Portugal Singapore
South Korea Switzerland Thailand Turkey Ukraine Vietnam

Oxford is a registered trade mark of Oxford University Press
in the UK and in certain other countries

Published in the United States
by Oxford University Press Inc., New York

© Oxford University Press 2009

The moral rights of the author have been asserted
Database right Oxford University Press (maker)

First Published 2009

All rights reserved. No part of this publication may be reproduced,
stored in a retrieval system, or transmitted, in any form or by any means,
without the prior permission in writing of Oxford University Press,
or as expressly permitted by law, or under terms agreed with the appropriate
reprographics rights organization. Enquiries concerning reproduction
outside the scope of the above should be sent to the Rights Department,
Oxford University Press, at the address above

You must not circulate this book in any other binding or cover
and you must impose the same condition on any acquirer

British Library Cataloguing in Publication Data
Data available

Library of Congress Cataloging in Publication Data
Field, Timothy R.
Electromagnetic scattering from random media/Timothy R. Field.
p. cm — (International series of monographs on physics; no. 144)
ISBN 978–0–19–857077–6
1. Stochastic processes. 2. Random fields. 3. Mathematical physics. 4. Electromagnetic
waves–Scattering. I. Title.
QC20.7.S8.F54 2008
530.14′1—dc22   2008042115

Typeset by Newgen Imaging Systems (P) Ltd., Chennai, India
Printed in the UK
on acid-free paper by
the MPG Books Group

ISBN 978–0–19–857077–6

1 3 5 7 9 10 8 6 4 2

To my family

PREFACE

The topic covered by the monograph is the scattering of electromagnetic waves from random media. The book falls under the general categories of Mathematical Physics and Stochastic Methods. The treatment is based upon recent developments in the journal literature by the author, which is provided here with sufficient background material required for a self-contained account.

The overall objective of the book is to formulate the space-time dynamics of electromagnetic scattering processes from random media in terms of stochastic differential equations (SDEs), and to demonstrate the utility of this theoretical framework in model simulation and experimental data analysis. The results provide various means for detecting anomalous behaviour in the scattering process, which should have significant consequences in several important application areas, including radar signal processing, optical scintillation effects and laser propagation through turbulent media.

The book is divided into three main parts: Part I – *Stochastic calculus* the opening chapters of which cover the necessary background to the Ito stochastic calculus, including the Ito stochastic differential and stochastic integral and related concepts. Part II – *Stochastic dynamics of the scattering process* discusses Rayleigh scattering, population models, scattering generating K-distributed noise resulting from the birth–death–immigration (BDI) population model and scattering from more general populations. SDEs for the various scattering processes are derived from first principles and their statistical and correlation properties are analysed in detail. This part contains the main body of results of the book, which constitute a space-time dynamical model for the scattering process. SDEs Part III – *Simulation and experiment* describes how the SDEs for the various scattering processes can be simulated in discrete time, and provides iterative schemes for reducing the error in this procedure. The theory is substantiated through experimental examples of scattering at optical and radar wavelengths, including laser propagation through a random phase screen, and anomaly detection for coherent radar scattering. The treatment is self-contained so as to be accessible by those familiar with elementary probability theory and electromagnetism. The book addresses, for the first time, the fundamental issue of the space-time dynamics of electromagnetic-scattering processes from random media. As such, it represents a significant advance in this area of research, and should prove indispensable to researchers and practitioners in this field at the postgraduate level and above. The statistical description of electromagnetic scattering from random media is important in a number of contexts. These include, for example, radar scattering, sea clutter suppression, and the propagation of laser light through turbulent atmospheric conditions. An understanding of the statistical properties of such scattering processes enables their characterization in terms of correlation functions and higher order statistics of various spatial and temporal physical scattering data. Under certain assumptions concerning the nature of the scattering population, it is possible to deduce that the scattered field intensity consists of two components, a Rayleigh or Gaussian speckle pattern, modulated by the local scattering cross-section, which fluctuate over two independent correlation timescales. This is the so-called compound K-distribution model, which has been the extant model for sea-surface radar scattering since its inception over two decades ago. Since then, although a prominent and intensive area of research, the problem of understanding the dynamics of the scattering process has remained unsolved. This problem is of vital significance, since a description of the scattering dynamics (in terms of propagators) implies the form of all correlation functions and higher order statistics, and provides a valid prior for the power spectral density.[1] An appropriate dynamical model of these scattering processes has recently been discovered. From very primitive physical assumptions it has been possible to provide a detailed description of the stochastic dynamics of the electromagnetic scattering process that captures all the physical degrees of freedom in the problem. This discovery and its implications for modelling electromagnetic scattering and propagation phenomena are the basis of the current monograph.

The major features of the book can be summarized as follows: this pioneering study describes electromagnetic scattering from a population whose size is fluctuating in space and time. Its key findings are to characterize the dynamical laws governing the time evolution of the scattered electromagnetic field; the physical model enables extraction of all spatio-temporal correlation information and higher order statistics; this enables the results of radar and laser scattering experiments to be interpreted, and has implications for real time anomaly detection; this work is significant also because it illustrates how ideas in the Black–Scholes theory of financial option pricing can be applied to a physical problem in which non-Gaussian noise processes play an essential role, revealing scope for cross-fertilization between these disparate areas.

The primary readership for the book is intended as researchers in academia and industry, in the areas of electromagnetic scattering, SDEs, diffusion modelling, radar systems engineering, signal processing, applications of non-Gaussian noise processes, quantum optics, laser propagation, population processes, radar systems engineering, stochastic volatility models in the context of financial mathematics, and those with a general interest in applications of stochastic methods to physics. The book should also benefit advanced graduate students working in these areas. Practitioners in these fields should find the book a useful resource in the design and implementation of the various models in both simulation and experimental data analysis. Lastly, we have attempted to access a broad readership

---

[1] The reader should compare the remarks in Jakeman and Tough (1988), p. 508: "A full analysis of the temporal correlation properties of the variables x and z implicit in (5.24) would require knowledge of its fundamental solution or propagator, which is, as yet, unknown. ..." *et seq.*

by making, to some extent, the various chapters self-contained so that readers with specific interests should be able to focus on those relevant parts of the text.

An alternative, and well developed, approach to the statistical description of scattered radiation, is based on a more detailed description of both the scattering medium and its interaction with the electromagnetic field. Much of this work has been motivated by problems in microwave and optical sensing and communication. The Russian contribution to this subject has been significant and is reviewed in detail in the following references: V.I. Tatarskii *'Wave propagation in a Turbulent Medium'*, McGraw-Hill, 1961; S.M. Rytov, Yu. A. Kravtsov, and V.I. Tatarskii, *'Principles of Statistical Radiophysics'*, Vols 1–4, Springer Verlag, 1986; F.G. Bass and I.M. Fuks, *'Wave Scattering from Statistically Rough Surfaces'*, Pergamon Press, 1979; B.R. Levin, *'Fondements Theoretique de la Radio technique Statistique'*, MIR, Moscow Vols 1–2, 1973; Vol. 3, 1979. This approach has been extended, with particular attention being paid to the non-Gaussian statistical properties of the scattered radiation, by Jakeman and others in the UK. The recent book E. Jakeman and K. Ridley *'Modelling the Fluctuations in Scattered Waves'* describes these developments in authoritative detail. An extensive review, which also contains some discussion of the relationship of this physical model based approach to an analysis based on SDEs, is provided by E. Jakeman and R.J.A. Tough, 'Non-Gaussian models for the statistics of scattered waves.' *Adv. Phys.* **37**, 471, 1988.

The idea for writing the book arose from a suggestion of Simon Haykin, to whom we express sincere appreciation, following a two-day series of invited lectures by the author at McMaster University in October 2002.

T. R. F.
June 2007

ACKNOWLEDGEMENTS

We are grateful for academic support from the Departments of Electrical and Computer Engineering and Mathematics at McMaster University and the Brain Body Institute. Amongst those colleagues who deserve special thanks are John Bienenstock, Alexander Bain, Simon Haykin FRSC, Thomas Hurd, Eric Jakeman FRS, Vikram Krishnamurthy, John McWhirter FRS, Jose Principe, David Sherrington FRS, and Kon (Max) Wong FRSC, and my graduate students Patrick Fayard and Tao (Stephen) Feng. We especially thank Robert Tough for collaboration at the early stages and inspiring much of this scientific development, and for contributing the majority of Chapter 11. His constant mentoring and encouragement has been an essential part of the success of establishing this research area. We also thank former colleagues Brian Bramson, Richard Glendinning, Stephen Luttrell, and John O'Loghlen for valuable discussions, and Samantha Lycett and Kevin Ridley for supplying experimental data studied in Chapter 12.

The author acknowledges the award of a Discovery Grant from the Natural Sciences and Engineering Research Council of Canada.

CONTENTS

I STOCHASTIC CALCULUS

1 Heat equation and Brownian motion — 3
  1.1 Einstein's 1905 derivation — 3
    1.1.1 Green's function solution — 5
  1.2 Brownian motion — 5

2 Ito calculus — 7
  2.1 Ito stochastic integral — 7
    2.1.1 Ito isometry — 9
  2.2 Ito differential — 9
    2.2.1 Ito's formula — 10
    2.2.2 Ito product rule — 10

3 Stochastic differential geometry — 12
  3.1 Diffusions on manifolds — 12
  3.2 Kinematics of diffusion — 16

4 Examples of stochastic differential equations — 20
  4.1 Geometric Brownian motion — 21
  4.2 Bessel process — 21
  4.3 Solution to Laplace's equation — 23

II DYNAMICS OF THE SCATTERING PROCESS

5 Diffusion models of scattering — 27
  5.1 Non-Gaussian statistical models — 28
  5.2 Dynamics of the vector scattering process — 30
    5.2.1 Natural boundaries — 32
  5.3 Correlation in the vector scattering process — 35

6 Rayleigh scattering — 40
  6.1 Quadrature components — 40
    6.1.1 Rayleigh intensity — 41
    6.1.2 Statistical properties — 42
  6.2 Random walk model — 43
    6.2.1 Ornstein–Uhlenbeck process — 44

7 Population dynamics — 46
  7.1 Master equations and the Kramers–Moyal expansion — 46
  7.2 Birth–death–immigration processes — 47
  7.3 Continuum diffusion limit — 48

8	**Dynamics of K-scattering**	52
	8.1 Stochastic dynamics of K-amplitude process	53
	8.1.1 Intensity	55
	8.1.2 Phase	57
	8.2 Geometry of K-amplitude fluctuations	59
	8.3 Asymptotic behaviour	60
	8.3.1 Equilibrium distribution	60
	8.3.2 Detailed balance	62
	8.4 Correlation and spectra	63
	8.4.1 Intensity autocorrelation	63
	8.4.2 Power spectral density	64
	8.5 Interpretation and implications	67
9	**Models of weak scattering**	69
	9.1 Weak scattering amplitudes	70
	9.2 Stochastic dynamics	71
	9.3 Geometry of amplitude fluctuations	74
	9.4 Asymptotic behaviour	79
	9.4.1 Detailed balance	85
10	**Scattering from general populations**	88
	10.1 Extended random walk model	88
	10.2 Generalized dynamics	91

III SIMULATION AND EXPERIMENT

11	**Simulation of K-scattering**	99
	11.1 Iterative schemes	99
	11.2 Rayleigh and gamma processes	105
	11.3 Compound K-model	109
	11.4 Influence of Doppler on volatilities	111
	11.5 Coherent clutter	113
	11.5.1 Target returns in clutter	114
	11.6 Second-order algorithms	115
12	**Experimental tests**	124
	12.1 Scattering at optical wavelength	125
	12.1.1 Random phase screen	125
	12.2 Scattering at radar wavelength	126
	12.2.1 Coherent sea clutter	127
	12.2.2 Anomaly detection	132
13	**Non-linear dynamics of sea clutter**	138
	13.1 Hybrid AM/FM model of sea clutter	140
	13.2 Non-linear dynamics from SDE theory	141
	13.3 Radar parameters	144

	13.3.1 Superposition	144
	13.3.2 Sea state and polarization	145
14	**Observability of scattering cross-section**	**150**
	14.1 Simulated data	150
	14.2 Experimental applications	151
A	**Stability and infinite divisibility**	**153**
B	**Ito versus Stratonovich stochastic integrals**	**155**
C	**Filtrations, conditional probability, and Markov property**	**158**
D	**Girsanov's theorem**	**159**
	D.0.1 Relation to mathematical finance	160
E	**Partition function solution to BDI model**	**161**
F	**Summary of K-scattering**	**165**
	F.1 Rayleigh scattering	165
	F.2 K-distributed noise	166
	F.2.1 Amplitude	167
	F.2.2 Intensity	167
	F.2.3 Phase	168
	F.2.4 Geometry of fluctuations	168
G	**Iterative solution for vector processes**	**170**
H	**Open problems**	**172**
I	**Suggested further reading**	**174**
References		**177**
Index		**181**

Part I

Stochastic calculus

The part introduces the concepts and techniques of stochastic calculus that enable us to describe dynamical (scattering) processes in terms of stochastic differential equations (SDEs). Chapter 1 begins with an historical account of the heat (diffusion) equation in terms of Einstein's 1905 derivation and the corresponding Green's function solution. This leads naturally to the concept of the Wiener process, which describes the trajectory of an individual particle amongst a cloud whose density evolves according to the heat equation. Equipped with the Wiener process as an essential ingredient for the random driving force, in Chapter 2 we introduce the reader to the basic concepts and techniques of the Ito (and, to some extent, Stratonovich) stochastic calculus that we shall require in the description of scattering, and highlight its key aspects that differ from notions of the classical calculus.

Since we shall be concerned with physical scattering processes that are represented as vectors, and also in part for mathematical edification, we develop the subject further from the point of view of stochastic differential geometry in Chapter 3. The part concludes in Chapter 4 with an account of some specific examples of SDEs and their applications, which serve to illustrate the various concepts and techniques of the stochastic calculus introduced in the previous two chapters.

1

HEAT EQUATION AND BROWNIAN MOTION

This chapter explores the interrelationship between the theory of the heat process, in terms of the probability density function for a cloud of diffusing particles, and the underlying stochastic process pertaining to individual particles, which generates such an ensemble average behaviour that is called Brownian motion.

1.1 Einstein's 1905 derivation

Of considerable historical interest and significance is Albert Einstein's famous original 1905 derivation of the heat equation from considerations of Brownian motion (Einstein 1905), which we shall recall here. The argument, which is elegant yet mathematically relatively unsophisticated, contains many of the seeds of modern stochastic calculus.[2]

We consider a cloud of particles (pollen grains) suspended in a (stationary) fluid that undergo random 'thermal' motion due to the impacts they experience with molecules in the fluid. Let $\rho(x,t)$ denote the density of the Brownian particles at location x and time t. Suppose we consider an individual particle and the displacements Δ it experiences in some small time τ, and let the probability density of Δ be the function $\phi(\Delta)$ which is assumed to have the following two properties: 1) ϕ is a symmetric function, 2) ϕ decays very rapidly for $|\Delta| > \epsilon$, $0 < \epsilon \ll 1$. The first property arises since a particle without 'drift' is equally likely to be displaced in either direction, while the second is due to the very small chance that a particle experiences a large displacement in a short time. Now, the number of particles lying in the interval $(x, x + dx)$, which experience displacement Δ is

$$dn = n\phi(\Delta)d\Delta. \tag{1.1}$$

The heat equation concerns the evolution of the density ρ in space and time. To this end we write

$$\rho(x, t + \tau) = \rho(x, t) + \tau \frac{\partial \rho}{\partial t} + o(\tau). \tag{1.2}$$

On the other hand, we can express the left-hand side as an integral over space, via the following lucid reasoning. The events that a Brownian particle finds itself

---

[2] The reader may also be interested to compare the (broadly) contemporaneous work of Louis Bachelier on random walks, in the context of finance, which appears in Bachelier's doctoral thesis (*orig.* Bachelier 1900) and thus predates Einstein's famous 1905 paper on the Brownian movement.

at location x at time $t + \tau$ is logically equivalent to the event that it is location $x - \Delta$ at the earlier time t and experiences displacement Δ in the subsequent time τ, for all possible choices of Δ. This can be represented mathematically as the relation

$$\rho(x, t + \tau) = \int_{-\infty}^{\infty} \rho(x - \Delta, t) \phi(\Delta) \, d\Delta, \tag{1.3}$$

where, to be precise, we have used the following three facts: any equivalent pair of events, such as those identified above, have equal probabilities; distinct choices of Δ yield a set of disjoint events and so the probability of their union is additive (reflected in integral on the right-hand side above); the displacements are independent of space-time location, and so the joint probability of these attributes factorizes. Now, equating the two expressions for $\rho(x, t + \tau)$ above is the essence of Einstein's argument. Expanding (1.3) in powers of Δ we obtain

$$\rho(x, t + \tau) = \int_{-\infty}^{\infty} \left\{ \rho - \Delta \frac{\partial \rho}{\partial x} + \frac{1}{2} \Delta^2 \frac{\partial^2 \rho}{\partial x^2} + o(\Delta^2) \right\} \Big|_{(x,t)} \phi(\Delta) \, d\Delta. \tag{1.4}$$

Terms of $o(\Delta^2)$ can be neglected since we assume that ϕ decays rapidly for $|\Delta| > \epsilon$, $\epsilon \ll 1$ and so $\int \Delta^n \phi \approx 0$ (since $\Delta^n \approx 0$ for $|\Delta| < \epsilon$), $n > 2$. Also, $\int \Delta \phi$ vanishes since ϕ is assumed to be an even function, and since it is a probability density function $\int \phi = 1$. Thus we are left with

$$\rho(x, t + \tau) = \rho(x, t) + \tau D \frac{\partial^2 \rho}{\partial x^2} \tag{1.5}$$

in which we define the *diffusion coefficient* D as

$$D := \lim_{\tau \to 0} \frac{1}{2\tau} \int \Delta^2 \phi. \tag{1.6}$$

Remark. In this derivation of the diffusion coefficient we can see both the mathematical and physical origins of the identity for the quadratic variation of the Wiener process, namely the non-vanishing integral of $dW_t^2 = dt$, that is familiar in the modern Ito calculus.

Now comparing (1.5) with (1.2) and taking the limit as $\tau \to 0$ we obtain the *heat equation*

$$\frac{\partial \rho}{\partial t} = D \frac{\partial^2 \rho}{\partial x^2}. \tag{1.7}$$

A generalization of Einstein's argument that incorporates a drift (corresponding to a displacement density ϕ that is non-symmetric) and retains all higher derivatives in space is the *Kramers–Moyal* expansion (see e.g. Risken 1989) that we shall encounter later in the context of population models in Section 7.1.

1.1.1 Green's function solution

The fundamental Green's function solution to this equation, i.e. the solution to the initial value problem $\rho(x,0) = \delta(x)$ (where δ is a Dirac delta function) is the familiar Gaussian

$$\rho(x,t) = \frac{1}{\sqrt{4\pi Dt}} \exp\left(\frac{-x^2}{4Dt}\right) \qquad (1.8)$$

which has the variance property

$$\text{Var}_t[x] = 2Dt. \qquad (1.9)$$

This corresponds to an individual particle motion following a stochastic process X_t with stochastic differential equation (SDE):

$$dX_t = \sqrt{2D}\,dW_t \qquad (1.10)$$

in which W_t is the familiar Wiener process.

The reader should also compare the *Remarks on time symmetry* below in relation to the discussion of the current and osmotic velocities for a pure heat process.

Finally, we remark that it is intriguing from a historical perspective that the heat equation exhibits *non-causal* propagation (as can be seen explicitly through the Green's function solution provided in (1.8), observing that ρ is positive for arbitrarily small times at arbitrarily large distances). It is therefore incompatible with the advent of Einstein's special theory of relativity that occurred in the same year 1905, Einstein's *Annus Mirabilis*.

1.2 Brownian motion

A Brownian motion or Wiener process W_t is defined, consistently with Einstein's considerations above, as that which satisfies the independent increments property

$$\mathbf{E}\left[(W_v - W_u)(W_t - W_s)\right] = 0, \qquad (1.11)$$

where $s < t \leq u < v$ (so the intervals (s,t), (u,v) are non-overlapping) and such that the distribution of $W_t - W_s$ is the Gaussian $\mathcal{N}(0, |t-s|)$, and is subject to the initialization $W_0 = 0$.

The autocorrelation function of the Wiener process $\mathbf{E}[W_s W_t]$, $s \leq t$, can be calculated by writing the second factor under the expectation as $(W_t - W_s) + W_s$ and then applying the above independent increments property, which leads to

$$\mathbf{E}[W_s W_t] = s. \qquad (1.12)$$

Therefore, the Wiener process, although it has constant mean zero, fails to be wide sense stationary.

This process can also be constructed from the continuum limit of a discrete random walk, as follows. Consider a discrete random walk in discrete time, starting at the origin, with step size ϵ and time step δt. Suppose that the displacement at each step is drawn from the discrete set $\{\epsilon, -\epsilon\}$ with equal probability. Let the step size and time step scale relative to each other according to (cf. the discussion of Einstein's derivation above) $\epsilon^2 = \delta t$, for all $\delta t > 0$. Then, from the central limit theorem applied to the displacement after n steps in the interval $[0, t]$, where t is fixed and $n \to \infty$, $\delta t \to 0$, we find that this is $\mathcal{N}(0, t)$ distributed and, by construction, satisfies the independent increments property. The process so constructed therefore coincides (in probability) with the above Wiener process defined in the standard (continuum) manner.

2
ITO CALCULUS

This chapter develops the basic tools of the Ito calculus that flow from the development of the Brownian motion or Wiener process in Chapter 1. We emphasize those particular techniques that are relevant to the detailed analysis of the scattering processes encountered in the chapters that follow.

2.1 Ito stochastic integral

Recall, from the discussion of Brownian motion in the previous chapter, that a Wiener process W_t is defined as a continuous time stochastic process which satisfies the following two properties: first, the *independent increments property*

$$\mathbf{E}\left[(W_v - W_u)(W_t - W_s)\right] = 0, \tag{2.1}$$

where $s < t \leq u < v$, so the intervals (s,t) and (u,v) are non-overlapping; second, the distribution of the increment $W_t - W_s$ is the Gaussian with mean zero and variance $|t - s|$, i.e.,

$$W_t - W_s \sim \mathcal{N}(0, |t - s|); \tag{2.2}$$

third, the process is defined subject to the initialization $W_0 = 0$.

Equipped with this Wiener process, one is able to extend the ordinary motions of classical calculus to include a certain family of stochastic integrals, which differ from the ordinary Riemannian integral in that the integration measure becomes a *random* quantity, the increment in the Wiener process. Thus, in place of the standard Riemannian integral,

$$\int_0^t f(s)\,\mathrm{d}s \tag{2.3}$$

for some integrable function $f(s)$ (e.g. a continuous one), we shall consider replacing the measure $\mathrm{d}s$ by the stochastic differential $\mathrm{d}W_s$. In general, the integrand will be considered to be a stochastic process σ_s, the so-called (stochastic) *volatility*. As we shall see, in the sections that immediately follow, this concept of stochastic integration leads to corresponding modifications in the classical notion of derivative, that will be quite essential to our development of scattering dynamics.

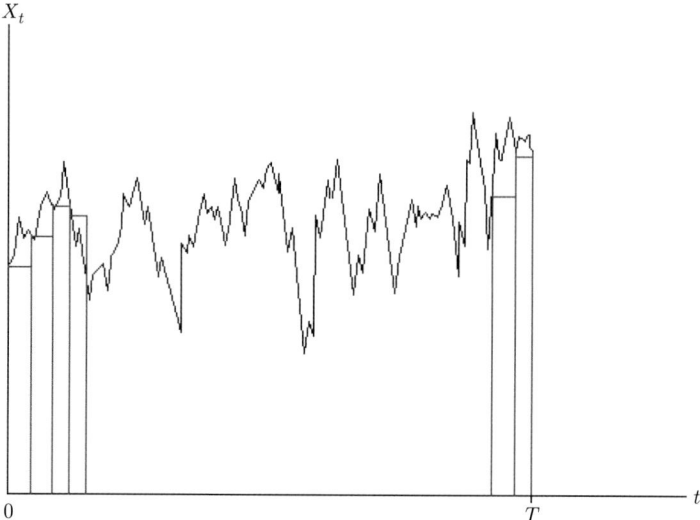

FIG. 2.1. Ito stochastic integral – the volatility σ_s is evaluated at the extreme *left* point of each subinterval, before multiplying by the Wiener measure dW_s.

Now, in light of the above expository remarks, a stochastic integral is defined as a quantity of the form

$$\int_a^b \sigma_s \, dW_s, \qquad (2.4)$$

wherein the *measure* of integration (the increment of the Wiener process) is itself *random*. Such an integral differs in nature from the classical Riemann integral, and is illustrated in Fig. 2.1. It is important (unlike in the case of the classical Riemannian integral, where it is irrelevant) that in the discrete sum approximating the Ito integral, the integrand is evaluated at the *left*-most point of each subinterval. Thus, in the Ito interpretation,

$$\int_a^b \sigma_s dW_s = \lim_{\delta s \to 0} \sum_{j=0}^{j_{\max}} \sigma_{a+j\delta s}(W_{a+(j+1)\delta s} - W_{a+j\delta s}), \qquad (2.5)$$

where $j_{\max} = (b-a)/\delta s - 1$. A detailed comparison of the above Ito integral and the (intimately related) Stratonovich integral, which evaluates the integrand at the *midpoint* of each subinterval, is provided in Appendix B.

Remarks. *On expectations of stochastic integrals.* Owing to the distributive property of the expectation functional $\mathbf{E}$ over addition, the definition (2.5), and the symmetry of the fluctuations δW_t with respect to sign (cf. the associated symmetry of the displacement distribution function $\phi(\Delta)$ discussed in Section 1.1), the expectation of a stochastic integral is *zero*, i.e.

$$\mathbf{E}\left[\int_a^b \sigma_s \mathrm{d}W_s\right] = 0. \tag{2.6}$$

In comparison with the above remarks, an important related property of such integrals, when they arise as products, that we shall require is the following.

2.1.1 Ito isometry

The *Ito isometry* expresses the expectation of a product of a pair of stochastic integrals as an ordinary 1-dimensional integral, as follows:

$$\mathbf{E}\left[\left(\int_0^t \phi_s \mathrm{d}W_s\right)\left(\int_0^t \psi_u \mathrm{d}W_u\right)\right] = \int_0^t \mathbf{E}\left[\phi_s \psi_s\right] \mathrm{d}s. \tag{2.7}$$

Thus, in terms of distribution valued functions the expected product stochastic differential $\mathbf{E}\left[\mathrm{d}W_s \mathrm{d}W_u\right]$ behaves like a Dirac delta function weighted by the product time measure, $\delta(s-u)\mathrm{d}s\mathrm{d}u$. This feature is a reflection of the independent increments property of the Wiener process together with the familiar identity for the quadratic variation $\mathrm{d}W_t^2 = \mathrm{d}t$.

Remarks. *On products of stochastic differentials.* It is worth remarking further in this type of context on two other instances of product differential relations for the Wiener process – this shall avoid confusion and also highlight some of the care that is needed in manipulating such expressions. For a pair of uncorrelated Wiener processes W and $\tilde{W}$ it is consistent to write $\mathrm{d}W_s \mathrm{d}\tilde{W}_s = 0$ (differentials with equal time indices) *without* the need for expectation, since this quantity equals zero upon integration.[3]

However, it is not consistent to set the product stochastic differential $\mathrm{d}\xi_s \mathrm{d}\tilde{\xi}_u$ (with *distinct* time indices) equal to zero if ξ, $\tilde{\xi}$ are independent – this fact is evident from considering the associated double stochastic integral

$$\iint_0^t \phi_s \psi_u \mathrm{d}\xi_s \mathrm{d}\tilde{\xi}_u = \left(\int_0^t \phi_s \mathrm{d}\xi_s\right)\left(\int_0^t \psi_u \mathrm{d}\tilde{\xi}_u\right) \tag{2.8}$$

which does *not* vanish.

2.2 Ito differential

According to the prescription for taking the Ito stochastic integral in Section 2.1, we consider a (vector-valued) stochastic (Ito) process X_t^i with differential

---

[3] Consider the integral as the limit of the discrete sum $\sum_{i=1}^N n_{s_i} \tilde{n}_{s_i} \delta s$ where $t = N\delta s$ and $N \to \infty$ and n, $\tilde{n}$ are $\mathcal{N}(0,1)$ random variables arising in $\delta W_s = n_s \delta s^{1/2}$ – the limit then yields the mean of the random variable $n\tilde{n}$ (scaled by t), which vanishes. If this product differential is considered against a continuous ($\mathcal{F}_s$-adapted) process ϕ_s then consider $\lim_{\Delta \to 0} \int \phi^{(\Delta)} \mathrm{d}W_s \mathrm{d}\tilde{W}_s$ where $\phi^{(\Delta)}$ is an approximation to ϕ that is piecewise constant over successive sub-intervals of length Δ, and apply the preceding argument (to each subinterval).

$$dX_t^i = b^i dt + d\xi_t^i \tag{2.9}$$

in which ξ_t satisfies $d\xi_t^i d\xi_t^j = \sigma^{ij} dt$. The meaning of this 'differential' statement, which is infinitesimal in nature, can be understood as a mathematical shorthand for a corresponding integral equation, in which the integrals that arise, where relevant, are understood in the Ito sense, as explained in Section 2.1.

2.2.1 Ito's formula

It is often natural and relevant to consider prescribed functions of a stochastic process such as in (2.9), together with their stochastic differentials. Accordingly, let us define a new related stochastic process

$$F_t := f(X_t^i, t) \tag{2.10}$$

wherein the function f is known. For example, in the description of random phasor dynamics in Chapter 6, it is the complex exponential function that is relevant. *Ito's formula* then provides us with a convenient explicit expression for the stochastic differential of F_t in terms of f and dX_t. The key to the result, which constitutes a departure from the classical rules of calculus, is that the underlying process X_t has positive quadratic variation, i.e. that its *squared* fluctuations integrate to a non-zero amount. Consistently, *retaining* second-order terms, we write

$$dF_t = \frac{\partial f}{\partial t} dt + \frac{\partial f}{\partial X^i} dX_t^i + \frac{1}{2} \frac{\partial^2 f}{\partial X^i \partial X^j} dX_t^i dX_t^j + o(dt). \tag{2.11}$$

Substituting (2.9) into the above we deduce that

$$dF_t = \left(\frac{\partial f}{\partial t} + b^i \frac{\partial f}{\partial X^i} + \frac{1}{2} \sigma^{ij} \frac{\partial^2 f}{\partial X^i \partial X^j} \right) dt + \frac{\partial f}{\partial X^i} d\xi_t^i \tag{2.12}$$

which statement is *Ito's formula*.

2.2.2 Ito product rule

The origin of the *Ito product rule* is similar in essence to the existence of a quadratic variation that enters into the Ito formula, where now we consider a pair of processes. Essentially, the result says that we must take into account product differentials when calculating the differential of a product, and thus the result constitutes a modification of the classical Leibnitz rule. Explicitly, for the product of a pair of stochastic processes $U_t V_t$ we write the incremental product

$$\delta(U_t V_t) = U_t \delta V_t + V_t \delta U_t + \delta U_t \delta V_t + o(\delta t) \tag{2.13}$$

which takes the infinitesimal form

$$d(U_t V_t) = U_t dV_t + V_t dU_t + dU_t dV_t \tag{2.14}$$

in which it is essential to observe that the third term on the right-hand side above is *non-zero*.

Remarks. Observe that in the product expression $\mathrm{d}U_t \mathrm{d}V_t = \mathbf{E}[\mathrm{d}U_t \mathrm{d}V_t]$, i.e. the expectation $\mathbf{E}$ is *superfluous*; in the case that the Wiener processes driving U and V are independent this product is zero, corresponding to a zero (U,V) entry in the diffusion tensor σ^{ij}.

3

STOCHASTIC DIFFERENTIAL GEOMETRY

The chapter further develops the theory of (non-degenerate) diffusion processes, considered as existing on an arbitrary manifold $\mathcal{M}$, from the point of view of stochastic differential geometry. In this description the diffusion tensor plays an essential underlying role in supplying the metrical structure of the manifold. We develop the kinematics of various quantities that arise naturally in this geometrical context, and explain their potential significance in a physical (scattering) context. This more abstract geometrical interpretation should prove illuminating for the pure mathematical reader, in terms of our treatment of the dynamics of the vector scattering processes encountered in subsequent chapters.

Our analysis is carried out using the natural metrical geometry supplied by the inverse of the (non-degenerate) contravariant diffusion tensor. The notion of stochastic velocity (and therefore acceleration) is not unique for a stochastic process, owing to the distinct notions of forward and backward drift velocities and time derivatives that are conditioned with respect to the past and future. In this respect the specification of velocity dynamics of a process requires some care, and we illustrate this concept with the heat process.

The chapter is organized as follows. We begin in Section 3.1 with an exposition of the elements of stochastic differential geometry for non-degenerate diffusion processes on a manifold $\mathcal{M}$, which are necessary for our discussion. Section 3.2 discusses the kinematics of diffusion and illustrates the distinction between the various stochastic notions of velocity, providing the heat process as a concrete example. The relation to the notion of detailed balance in this context is also explained, as it will be required in the description of complex-valued (vector) scattering processes. We explain the concept of a conserved fluid current in a stochastic setting, as distinct from the notion of forward drift in the Fokker–Planck equation (FPE).

The reader should be familiar with the basic elements of differential geometry, diffusion theory and the Ito stochastic calculus. Tensorial indices shown in roman bold type denote the coordinate basis representation, while those in plain italic are abstract indices that serve to indicate to which space the relevant geometrical object belongs. We adopt the Einstein summation convention throughout.

3.1 Diffusions on manifolds

We introduce the diffusion process X_t^i on a manifold $\mathcal{M}$ and the associated contravariant diffusion tensor σ^{ij}, which in the non-degenerate case has inverse σ_{ij} that provides a natural metrical structure on $\mathcal{M}$. The Levi-Civita connection of this metric is introduced for the purpose of differential geometric operations, and has the simplifying property that it annihilates the diffusion tensor. In this geometry it is natural to work with the Ito drift of the process, whose transformation properties are tensorial with respect to $(\mathcal{M}, \sigma_{ij})$. The situation in this regard is contrasted with the case of the Kolmogoroff forward drift obtained by regarding the process as existing on $\mathbf{R}^n$ (Karatzas and Shreve 1988). The current treatment has the advantage over previous work (cf. Nelson 1967a, 1985) that each geometrical operation may be expressed in the coordinate free abstract index notation, as elucidated for example, in Penrose and Rindler (1984).

Consider a continuous time diffusion process X_t^i on a manifold $\mathcal{M}$ of dimension n taken to satisfy the (time independent) Ito stochastic differential equation (SDE)

$$\mathrm{d}X_t^\mathbf{i} = \beta^\mathbf{i}(X_t^i)\mathrm{d}t + \sum_j \sigma^\mathbf{i}_{(j)}(X_t^i)\mathrm{d}W_t^{(j)} \tag{3.1}$$

where $\{W_t^{(j)}\}$ are a collection of n independent Wiener processes.[4] The (contravariant) diffusion tensor $\sigma^{\mathbf{ij}}$ is then determined by $\mathrm{d}X_t^\mathbf{i}\mathrm{d}X_t^\mathbf{j} = \sigma^{\mathbf{ij}}\mathrm{d}t$, so that consistently $\sum_r \sigma^\mathbf{i}_{(r)}\sigma^\mathbf{j}_{(r)} = \sigma^{\mathbf{ij}}$. We distinguish between two notions of drift on the manifold as follows (cf. Nelson 1985). Regarding X_t^i as a process on a coordinate patch of $\mathbf{R}^n$ we have the Kolmogoroff mean forward drift (Karatzas and Shreve 1988) defined as

$$\beta^\mathbf{i} = \lim_{\delta t \to 0} \mathbf{E}_t\left[\frac{X_{t+\delta t}^\mathbf{i} - X_t^\mathbf{i}}{\delta t}\right] \tag{3.2}$$

and the corresponding mean backward drift

$$\tilde{\beta}^\mathbf{i} = \lim_{\delta t \to 0} \mathbf{E}_t\left[\frac{X_t^\mathbf{i} - X_{t-\delta t}^\mathbf{i}}{\delta t}\right] \tag{3.3}$$

for $\delta t > 0$ in a coordinate chart $\{x^i\}$ that is normal with respect to the Euclidean metrical geometry δ_{ij} of $\mathbf{R}^n$ (e.g. Nelson 1967b). Denoting the Levi-Civita connection of δ_{ij} by ∂_i the probability density $\rho_\mathcal{E}$, with respect to the Euclidean volume measure on $\mathbf{R}^n$, satisfies the FPE (e.g. Risken 1989)

$$\frac{\partial \rho_\mathcal{E}}{\partial t} + \frac{\partial}{\partial x^i}(\rho_\mathcal{E} \beta^i) = \frac{1}{2}\partial_i \partial_j(\sigma^{ij}\rho_\mathcal{E}) \tag{3.4}$$

and the corresponding equation for the backward drift

$$\frac{\partial \rho_\mathcal{E}}{\partial t} + \frac{\partial}{\partial x^i}(\rho_\mathcal{E} \tilde{\beta}^i) = -\frac{1}{2}\partial_i \partial_j(\sigma^{ij}\rho_\mathcal{E}). \tag{3.5}$$

Now, for a non-degenerate diffusion we write $\sigma^{ij}\sigma_{jk} = \delta^i_k$, with $\sigma := \det\{\sigma_{\mathbf{ij}}\} \neq 0$. In this way the inverse σ_{ij} supplies the natural metrical geometry of $\mathcal{M}$ from

---

[4]The status of the parenthesized index is conceptually distinct from that of a tensor or its coordinate representation, and the index serves merely to label an individual member amongst the collection of Wiener processes.

a diffusion point of view, in that no extra metrical ingredient is required in the construction of a diffusion on $\mathcal{M}$. In the following sections we shall work mainly with the Levi-Civita connection ∇_i of the diffusion 'metric' σ_{ij}, with associated components of the covariant derivative given by

$$\nabla_i V^j = \partial_i V^j + \Gamma^i_{jk} V^k \tag{3.6}$$

with the Christoffel symbols defined in the standard way as

$$\Gamma^i_{jk} = \frac{1}{2}\sigma^{ip}(\partial_j \sigma_{pk} + \partial_k \sigma_{pj} - \partial_p \sigma_{jk}). \tag{3.7}$$

Thus our chosen ∇_i annihilates the diffusion tensor, which shall be a useful property in the calculations that follow. In the spirit of the σ_{ij} metrical geometry it is more convenient to define the *Ito* forward and backward drifts in the same way as in (3.2), (3.3), except that $\{x^i\}$ now refers to a coordinate chart that is normal with respect to the Levi-Civita connection ∇_i of the σ_{ij} metrical geometry. This drift quantity has the virtue that its components change in a tensorial way under coordinate transformations. The probability density ρ (for the *same* process) with respect to the invariant σ volume measure $\sigma^{1/2} d^n x^i$ is related to the previous density, that is tied to the volume measure of the Euclidean metric, by

$$\rho_{\mathcal{E}} = \rho\sqrt{\sigma} \tag{3.8}$$

and satisfies the covariant FPEs

$$\begin{aligned}\frac{\partial \rho}{\partial t} + \nabla_i(\rho b^i) &= \frac{1}{2}\sigma^{ij}\nabla_i\nabla_j\rho \\ \frac{\partial \rho}{\partial t} + \nabla_i(\rho \tilde{b}^i) &= -\frac{1}{2}\sigma^{ij}\nabla_i\nabla_j\rho.\end{aligned} \tag{3.9}$$

Adding these two equations yields the *equation of continuity* familiar from fluid dynamics (e.g. Batchelor 1967)

$$\frac{\partial \rho}{\partial t} + \nabla_i(\rho v^i) = 0. \tag{3.10}$$

Observe that, since the connection ∇ is designed to annihilate σ, the positioning of the σ^{ij} term on the right-hand side above has no effect, unlike in (3.4), (3.5). This feature will prove useful in the calculations that follow. Comparison of the two forms of the FPE yields[5]

$$b^i = \beta^i - \frac{1}{2\sqrt{\sigma}}\partial_j(\sqrt{\sigma}\sigma^{ij}). \tag{3.11}$$

---

[5]The comparison in fact shows that the ρ-weighted left- and right-hand sides of the equation below have equal divergences. Since this holds for arbitrary ρ independently of b, β, σ the identity follows.

This relationship can be re-expressed in terms of the Christoffel symbol Γ as follows. From the identity $\Gamma^k_{kj} = \partial_j(\log\sqrt{\sigma})$ we obtain for the second term on the right-hand side

$$\frac{1}{\sqrt{\sigma}}\partial_j(\sqrt{\sigma}\sigma^{ij}) = \partial_j \sigma^{ij} + \sigma^{ij}\Gamma^k_{kj}. \qquad (3.12)$$

We then define the trace of the Christoffel symbol (on its lower indices)

$$\Gamma^i := \sigma^{jk}\Gamma^i_{jk} = -\partial_j \sigma^{ij} - \sigma^{ij}\Gamma^k_{kj} \qquad (3.13)$$

and thus (3.11) can be written as

$$b^i = \beta^i + \frac{1}{2}\Gamma^i \qquad (3.14)$$

(cf. Nelson 1967a, 1967b, 1985). We deduce the transformation law for β^i from the Ito equation (3.1) which, on setting $\mathcal{W}^i := \sum_j \sigma^i_{(j)} dW^{(j)}$, implies that

$$dX^{i'} = \frac{\partial x^{i'}}{\partial x^i}(\beta^i dt + d\mathcal{W}^i) + \frac{1}{2}\frac{\partial^2 x^{i'}}{\partial x^i \partial x^j}d\mathcal{W}^i d\mathcal{W}^j + o(dt). \qquad (3.15)$$

Taking $\mathbf{E}_t$ of this equation and using the definition (3.2), in general coordinates, yields

$$\beta^{i'} = \frac{\partial x^{i'}}{\partial x^i}\beta^i + \frac{1}{2}\sigma^{ij}\frac{\partial^2 x^{i'}}{\partial x^i \partial x^j}. \qquad (3.16)$$

Recalling the transformation for the components of the Christoffel symbol

$$\Gamma^{i'}_{j'k'} = P^{i'}_i P^j_{j'} P^k_{k'} \Gamma^i_{jk} - P^j_{j'} P^k_{k'} \partial_j P^{i'}_k, \qquad (3.17)$$

where $P^{i'}_i = \partial x^{i'}/\partial x^i$, it follows that

$$\Gamma^{i'} = \frac{\partial x^{i'}}{\partial x^i}\Gamma^i - \sigma^{jk}\frac{\partial^2 x^{i'}}{\partial x^j \partial x^k}. \qquad (3.18)$$

Hence the non-tensorial terms in the transformations for the components β^i and Γ^i cancel in (3.14), and therefore the components b^i transform *vectorially*. Observe that, for a non-degenerate diffusion on $\mathcal{M}$, a sufficient condition for the Kolmogoroff and Ito drifts to coincide at a given point P is that $\partial_i \sigma_{jk}$ (or equivalently Γ^i_{jk}) vanish there. The converse does not hold however, since it is possible for Γ^i to vanish with $\partial_i \sigma_{jk}$ not equivalently zero. By contrast, a similar type of argument leads to the result that the components of the volatility and diffusion tensor transform tensorially.

The situation in respect of the abstract index notation can be summarized as follows (cf. eqns 7.1, 7.2 in Hughston 1996). The components of the abstract covariant derivative $\nabla_b \xi^a$ of a vector ξ^a in the tangent space to $T\mathcal{M}$ can be expressed via

$$\delta_{\mathbf{b}}^b \delta_a^{\mathbf{a}} (\nabla_b \xi^a) = \frac{\partial \xi^{\mathbf{a}}}{\partial x^{\mathbf{b}}} + \Gamma_{\mathbf{bc}}^{\mathbf{a}} \xi^{\mathbf{c}}, \qquad (3.19)$$

where $\delta_{\mathbf{a}}^a$ is a coordinate basis on a patch of $\mathcal{M}$ and $\delta_a^{\mathbf{a}}$ its dual. The coordinate stochastic differential (3.1) can be expressed via (3.14) as

$$\mathrm{d}x^{\mathbf{a}} + \frac{1}{2} \Gamma_{\mathbf{bc}}^{\mathbf{a}} \sigma^{\mathbf{bc}} \mathrm{d}t - \sum_i \sigma_{(i)}^{\mathbf{a}} \mathrm{d}W^{(i)} = b^{\mathbf{a}} \mathrm{d}t. \qquad (3.20)$$

This enables us to introduce the abstract Ito differential according to

$$\mathrm{d}x^a = \delta_{\mathbf{a}}^a \left(\mathrm{d}x^{\mathbf{a}} + \frac{1}{2} \Gamma_{\mathbf{bc}}^{\mathbf{a}} \sigma^{\mathbf{bc}} \mathrm{d}t \right) \qquad (3.21)$$

and hence (3.1) becomes the abstract geometrical equation[6]

$$\mathrm{d}x^a = b^a \mathrm{d}t + \sum_i \sigma_{(i)}^a \mathrm{d}W^{(i)}. \qquad (3.22)$$

The vector $\beta^i = \delta_{\mathbf{i}}^i \beta^{\mathbf{i}}$ then becomes a genuine tensorial object on $\mathcal{M}$, but one whose definition via (3.2) is tied to the choice of a normal coordinate chart with respect to the Levi-Civita connection ∂_i of the Euclidean metric δ_{ij} on $\mathbf{R}^n$. Henceforth in abstract index expressions, indices will be raised and lowered with σ^{ij}, σ_{ij}, respectively in accordance with the Einstein summation convention.

This geometrical characterization of the diffusion tensor will be relevant to the treatment of the amplitude fluctuations for the various scattering processes we consider in Part III. In expressions that arise hereafter we shall use abstract index or coordinate basis notation as appropriate to the context and clarity of exposition, and the summation convention shall be assumed throughout.

3.2 Kinematics of diffusion

We begin by introducing some additional kinematic quantities that will be relevant to our treatment of scattering from a stochastic point of view. We define the forward and backward *generators* of the process $X_t^{\mathbf{i}}$ respectively via

$$\begin{cases} \mathcal{D}f(t,x) = \lim_{\delta t \to 0} \mathbf{E}_{t,x} \left[f(t+\delta t, X(t+\delta t)) - f(t, X(t)) \right] / \delta t \\ \tilde{\mathcal{D}}f(t,x) = \lim_{\delta t \to 0} \mathbf{E}_{t,x} \left[f(t, X(t)) - f(t-\delta t, X(t-\delta t)) \right] / \delta t \end{cases} \qquad (3.23)$$

for an arbitrary function $f(t,x)$. Since X_t is taken to satisfy (3.1) we deduce from Ito's formula applied to $f(t, X_t)$ that these equations may be expressed in

---

[6] We recommend that the interested mathematical reader consult Hughston (1996) for an eloquent and thorough exposition of the abstract index notation in the context of stochastic differential geometry.

operator form as

$$\mathcal{D} = \frac{\partial}{\partial t} + b^i \nabla_i + \frac{1}{2}\sigma^{ij}\nabla_i\nabla_j,$$
$$\tilde{\mathcal{D}} = \frac{\partial}{\partial t} + \tilde{b}^i \nabla_i - \frac{1}{2}\sigma^{ij}\nabla_i\nabla_j. \qquad (3.24)$$

We further define the 'current' and 'osmotic' velocities v^i, u^i respectively according to

$$v^i := \frac{1}{2}(b^i + \tilde{b}^i), \qquad (3.25)$$

$$u^i := \frac{1}{2}(b^i - \tilde{b}^i). \qquad (3.26)$$

To calculate these kinematical quantities for specific dynamical processes we shall require the *osmotic equation* (Nelson 1967a, 1985) which relates the osmotic velocity u^i to the probability density via

$$u^i = \frac{1}{2}\sigma^{ij}\nabla_j \log \rho. \qquad (3.27)$$

It is illuminating to study the situation for pure diffusion (the heat equation) in the context of these kinematical quantities. The pure diffusion process has vanishing (forward) drift, as we have seen from the elements of Einstein's derivation. Nevertheless, it has *non*-zero current, $v \neq 0$. This (stochastic) current corresponds to the intuitive notion of a 'flow' of diffusing particles which, as we shall see, is quite distinct from the notion of forward drift. The explicit form of this current can be derived from the Green's function solution (1.8) and the osmotic equation (3.27) in combination with the defining equation for the stochastic current (3.25). Since the drift vanishes, the current and osmotic velocities are equal and opposite. According to the osmotic equation and Green's function solution, we find

$$v = \frac{x}{2t}, \quad t > 0 \qquad (3.28)$$

which is notably independent of the diffusion coefficient D. Observe, from the above relation, that the signs of v and x coincide. Intuitively this is consistent – the expected location of a Brownian particle situated at $x > 0$ at some time t, at an *earlier* time, is *less* than x which spatial asymmetry stems from that in the prescribed initial condition. In other words, the particle is more likely to have come from the origin than (positive) infinity. The initial density (an infinite concentration at the origin) induces an essential spatial asymmetry with respect to the backward drift (cf. eqn (3.3)).[7]

Remarks. *On time symmetry.* Under 'time' reversal $\mathcal{T}: t \mapsto -t$ these quantities transform as

$$\begin{cases} \mathcal{T}[b^i] = -\tilde{b}^i \\ \mathcal{T}[\tilde{b}^i] = -b^i \end{cases} \tag{3.29}$$

and hence

$$\begin{cases} \mathcal{T}[u^i] = u^i \\ \mathcal{T}[v^i] = -v^i. \end{cases} \tag{3.30}$$

The reversal of v^i under $\mathcal{T}$ is in accordance with intuition since this velocity is identical to the Eulerian velocity field of fluid mechanics (Batchelor 1967), while the invariance of u^i under $\mathcal{T}$ is consistent with the osmotic equation (3.27). Similarly for the forward and backward generators we have

$$\mathcal{T}[\mathcal{D}] = -\tilde{\mathcal{D}}. \tag{3.31}$$

In addition, we observe that the pair of (covariant) FPEs (3.9) can be obtained from each other immediately by invoking the time-reversal operation $\mathcal{T}$ (since ρ, ∇, σ^{ij} are each invariant under this map, while $\partial/\partial t$ reverses sign).

Remarks. *On detailed balance.* In a general context, there exist two distinct notions of stochastic equilibrium, namely that of ordinary equilibrium and the stronger condition of detailed balance. Ordinary 'equilibrium' is attained when the probability density occurring in the FPE is time invariant. If, in addition, the stochastic current vanishes, then the condition of 'detailed balance' is said to hold. Clearly, from the above discussion, detailed balance implies equilibrium, but not conversely.

We return to examine such kinematical properties, in the context of detailed balance for specific scattering processes, in Sections 5.3, 8.3.2, and 9.4.1.

The interested reader may wish further to study the application of this stochastic differential geometry to the dynamics of quantum mechanical diffusion on general manifolds, as discussed in Field (2003). This paper explores non-degenerate diffusion processes on an arbitrary manifold, the dynamics of which arise from a principle of least action for a Lagrangian consisting of a kinetic term quadratic in the forward drift of the process and a local potential. The equation governing the action emerges as a stochastic Hamilton–Jacobi condition and is expressed in terms of the geometry determined by the Levi-Civita

---

[7] Note that, in the definition of the backward drift (3.3), the differentiation is still *forwards* in time, the distinction between the forward drift arising from the subtle difference in the positioning of the time increment with respect to the conditioning.

connection of the diffusion tensor. It is argued that there are essentially two dynamical structures for the rate of change of the drift in the presence of a local potential, consistent with the requirement of time-reversal symmetry. In both cases a conserved energy is identified. An alternative wave function and associated operator description reveal a complex structure in the dynamical equations, thus extending the earlier results of Edward Nelson on the stochastic treatment of the Schrödinger equation (Nelson 1967a, 1985).

4
EXAMPLES OF STOCHASTIC DIFFERENTIAL EQUATIONS

We begin the chapter with some general remarks concerning the physical motivation for the application of stochastic differential equation (SDE) theory to problems in the mathematical description of scattering from random media. Such scattering problems are characterized by two essential ingredients. First, the existence of a random population, whose size fluctuates in time. A measure of the size of this population (in the continuum limit) is what is referred to as the scattering cross-section and its statistical characteristics are independent of those of any external electromagnetic radiation that might be used to indirectly probe the properties of the population. From an experimental point of view, the cross-section is usually the object of primary interest and various experiments are designed to infer its behaviour from the properties of electromagnetic radiation that scatters from or propagates through the random medium.

Second, the electromagnetic component consists of a known transmitted wave that interacts with the random population, and is subsequently forward- or backscattered. The random elements that enter here are the fluctuations in mean power as determined by the cross-section, together with microscopic phase randomization effects that occur for each component of the scattered field that is essentially an electromagnetic property. The resulting electromagnetic energy therefore entails two degrees of randomness, inherited from the scattering population size and its microscopic details that affect the degree of coherence amongst the component scattered waves associated with each member of the population.

Thus, we are confronted with a coupled system that has an inherent two-fold randomness, namely, the scattering population and the associated electromagnetic radiation. Both of these components of the state of the system evolve temporally as well exhibiting spatial correlation. Thus, the scattering processes that we shall consider are fundamentally *continuous* time-random entities. It is therefore appropriate, and indeed necessary, to employ a stochastic methodology that is able to capture the dynamics of such processes in as complete a manner as possible, beyond eliminating the inherent randomness in the system.

From this point of view, conventional statistical descriptions, though they may be established and correct, do not represent a complete physical description, capturing only certain ensemble average characteristics, such as distribution, correlation, power spectra and certain higher order statistics. SDE theory in partnership with the Ito calculus, on the other hand, provides a rigourous mathematical framework that pertains to random states of the system that are separated in time on an *infinitesimal* scale. The theory is therefore able, in principle, to yield all higher order statistical properties of such a physical system.

The identification of the parameters in SDEs that make this possible, in this monograph, is achieved through both experimental and independent theoretical reasoning.

The following sections provide some common illustrative examples in SDE theory, so that the reader can become familiarized with the basic concepts and techniques that we shall require in the study of scattering processes.

4.1 Geometric Brownian motion

The following example is of particular interest in financial mathematics in modelling the behaviour of a stock price S_t, and its exact (random) solution illustrates the basic principles involved in applying Ito's formula.

We take the increment in S_t in proportion to the instantaneous value to behave according to the SDE

$$\frac{dS_t}{S_t} = \mu dt + \sigma dW_t. \qquad (4.1)$$

In the absence of the reciprocal term on the left-hand side, the solution entails a straightforward integration of the right-hand side, thus $S_T = \mu T + \sigma W_T$ (for $S_0 = 0$). For *geometric* Brownian motion, however, it is necessary to consider the stochastic differential of the logarithmic process which, via Ito's formula, has the SDE

$$d(\log S_t) = \frac{dS_t}{S_t} - \frac{dS_t^2}{2S_t^2} \qquad (4.2)$$

in which the novice to the Ito calculus should carefully observe the presence of the second-order term on the right-hand side. Then substituting (4.1) we find

$$d(\log S_t) = \left(\mu - \frac{1}{2}\sigma^2\right) dt + \sigma dW_t \qquad (4.3)$$

which has the merit of a pure differential on the left-hand side and eliminating the presence of S_t on the right-hand side. The exact (random) solution is therefore

$$S_T = S_0 \exp\left[\left(\mu - \frac{1}{2}\sigma^2\right) T + \sigma W_T\right]. \qquad (4.4)$$

Now, the exponent is Gaussian with mean $\left(\mu - \frac{1}{2}\sigma^2\right) T$ and variance $\sigma^2 T$. Hence S_T has *lognormal* behaviour, which example is of central importance in the Black–Scholes theory of option pricing (cf. Appendix D).

4.2 Bessel process

This example is a variant on the Wiener process in n-dimensions, with a natural geometrical interpretation, and which again illustrates the utility of Ito's formula.

In $\mathbf{R}^n$ the *Bessel process* is defined as the Euclidean metric distance of a n-dimensional Wiener process from the origin. Thus, if $\left\{W_t^{(i)}\right\}$ is a collection of independent Wiener processes, we define the Bessel process as

$$B_t^{(n)} := \sqrt{\sum_{i=1}^n W_t^{(i)2}} \qquad (4.5)$$

for integer $n > 1$. Our aim is now to calculate the SDE satisfied by $B_t^{(n)}$, according to the rules of Ito calculus. From Ito's formula we have

$$dB_t^{(n)} = \frac{1}{2}\left(\sum_{i=1}^n W_t^{(i)2}\right)^{-1/2}\left(\sum_{j=1}^n d\left(W_t^{(j)2}\right)\right)$$
$$- \left(\frac{1}{2}\right)^3\left(\sum_{i=1}^n W_t^{(i)2}\right)^{-3/2}\left[\sum_{j=1}^n d\left(W_t^{(j)2}\right)\right]^2. \qquad (4.6)$$

Now let us examine the term of the form $\sum d\left(W^2\right)$ occurring above in isolation. Via Ito's formula and the identity $dW^2 = dt$ we can express each term in the summation as

$$d\left(W_t^{(j)2}\right) = 2W_t^{(j)}dW_t^{(j)} + dt \qquad (4.7)$$

and thus

$$\sum_{j=1}^n d\left(W_t^{(j)2}\right) = 2\left(\sqrt{\sum_{i=1}^n W_t^{(i)2}}\right)d\tilde{W}_t + ndt \qquad (4.8)$$

in which we invoke a new (related) Wiener process $\tilde{W}_t$ (for $n > 1$).[8] Combining these results we deduce that the Bessel process satisfies the SDE

$$dB_t^{(n)} = \frac{1}{2B_t^{(n)}}(n-1)dt + d\tilde{W}_t \qquad (4.9)$$

for $B_t^{(n)} \neq 0$.[9]

Remarks. *Concerning positivity.* Directly from its definition, $B_t^{(n)}$ is everywhere positive (or zero); this is reflected in its SDE by the divergence of the drift as zero is approached, which property (in spite of the persistence of the Wiener fluctuating term) 'repels' the process from the origin. (The reader should compare the discussion of natural boundaries in Chapter 5.)

---

[8] The way in which the components combine to yield a new Wiener process arises from the stability of the Gaussian distribution, as discussed in Appendix A.

[9] For the trivial case $n = 1$ using the same techniques we find $dB_t^{(1)} = \text{sgn}(W_t)dW_t$ which is to be taken for $W <, > 0$ separately, thus maintaining positivity.

4.3 Solution to Laplace's equation

This (rather more abstract) example illustrates the intriguing connection between stochastic process theory and (elliptic) partial differential equations. We consider the solution to Laplace's equation

$$\nabla^2 \phi = 0 \qquad (4.10)$$

on some domain $\mathcal{D} \subset \mathbf{R}^n$, subject to the boundary condition $\phi|_{\partial \mathcal{D}} = f$ for some prescribed function f on the boundary $\partial \mathcal{D}$. This boundary value problem can be solved expediently via SDE theory (simulation), as follows.

We consider a pure Brownian motion X_t whose stochastic differential is simply that of the Wiener process, $\mathrm{d}W_t$, starting at an interior point $x \in \mathcal{D}$. Let this trajectory evolve from the initial time until it (first) reaches the boundary $\partial \mathcal{D}$[10] – denote this (random) 'hitting' point as $X_\tau \in \partial \mathcal{D}$ so that τ is the 'hitting time'. Then consider the expected value of the prescribed function f acquired at the hitting point, conditioned on the starting point x, i.e.

$$\phi(x) := \mathbf{E}^x \left[f(X_\tau) \right]. \qquad (4.11)$$

This is evidently a function of x and, since the generator of Brownian motion is the Laplacian (cf. the heat equation), the function ϕ so constructed satisfies Laplace's equation. Also, the boundary condition is manifestly satisfied since for $x \in \mathcal{D}$, we have $\tau = 0$ and $X_\tau = x$ (so $\mathbf{E}^x$ in (4.11) has no effect).

Similar constructions apply to the solution of Poisson's equation and other elliptic partial differential equations in relation to more general SDEs (see e.g. Oksendal 1998).

The theory of the *Ornstein–Uhlenbeck* process is of paramount importance in our treatment of scattering and we shall develop this separately in Chapter 6.

---

[10]The existence of a 'hitting' time at which this occurs is ensured by the continuity of the Brownian path.

Part II

Dynamics of the scattering process

In this part, we develop the detailed stochastic dynamics of the electromagnetic scattering from a random fluctuating population. We focus largely on the K-scattering model and its dynamical and statistical properties. Such a description extends naturally to processes of weak scattering, for which the scattered component is weak in comparison with an additional signal component that lies in coherent superposition.

In Chapter 5, we present a classification of continuous time diffusion processes for the local scattering cross-section and scattered intensity for an electromagnetic field scattering from a random medium, such that the asymptotic marginal distribution for the intensity is the K-distribution. These processes are represented as Ito stochastic differential equations (SDEs), which enable identification of their stochastic volatilities with certain free functions that serve to calibrate the model. Later, in Chapter 12, we shall use these results in a study of the volatility behaviour of the electromagnetic intensity scattered from a random phase screen, which establishes the form of the volatility function to a high accuracy. This model is then used to derive a SDE for the complex-valued amplitude, which is then applied to radar scattering from the sea surface for the purpose of anomaly detection. Chapter 6 develops, from first principles, the SDE pertinent to the case of Rayleigh scattering, for which the scattering cross-section is considered to be constant. In Chapter 7 we provide an exposition of the dynamics of fluctuating discrete populations which, in the limit of large population size, leads to a continuum population whose dynamics emerges in the form of an SDE for the scattering cross-section. Using this continuum population model, in combination with the description of the unit power Rayleigh process, in Chapter 8 we develop a complete dynamical description of the K-scattering complex amplitude process, its statistical properties and asymptotic behaviour. The geometrical structure of the infinitesimal fluctuations in the scattered amplitude are derived and interpreted physically. The effect of a Doppler term is considered and represented in terms of the amplitude SDE. The chapter concludes with an iterative (numerical) scheme for integration of the Rayleigh and gamma population constituents of the K-scattering process.

Chapter 9 develops the corresponding dynamical theory of weak scattering, which is designed to include the effect of a coherent offset contribution in the resultant amplitude with respect to which the scattered component is weak. This is seen to arise in essentially three distinct ways, namely Rice, homodyned K, and generalized K-scattering, depending on the manner in which the cross-section fluctuations are incorporated into the resultant amplitude. For each of these cases, we account for the form of the resultant amplitude in terms of a random walk model and provide the SDE dynamics. The corresponding geometry of the infinitesimal amplitude fluctuations and their interrelationship amongst the various cases is derived which, in the case of generalized K-scattering, exhibits a rich geometrical structure that we analyse in detail. The asymptotic statistical properties and detailed balance are derived in each case. The part concludes in Chapter 10 with some recent results concerning scattering from general populations in which we indicate how to obtain sharp statistical inference of the scattering cross-section via high-frequency sampling as developed later, in terms of simulated data, in Chapter 14.

It is worth clarifying some notational issues at this point. We adopt the consistent notation throughout for a continuous time stochastic processes q_t, with Ito differential $\mathrm{d}q_t$, We shall consistently adopt the notation for the decomposition of a general Ito process q_t into drift and volatility terms, respectively, as $\mathrm{d}q_t = b_t^{(q)}\mathrm{d}t + \sigma_t^{(q)}\mathrm{d}W_t^{(q)}$, with respect to some (fixed) probability measure on the space of paths, and define the diffusion coefficients $\Sigma^{(\cdot)}$ by $\mathrm{d}q_t \mathrm{d}p_t = \Sigma_t^{(q,p)}\mathrm{d}t$ and abbreviate the squared volatility via $\mathrm{d}q_t^2 = \Sigma_t^{(q)}\mathrm{d}t$. (The notation $\mathrm{d}q_t^2$, means that the (stochastic) differential 'd' is taken *before* its square.) The error surface $\mathcal{S}^q$ of a vector process q^i is then defined by the (inverse) quadratic form relation $\Sigma^{(q)-1}(\delta q) = 1$. We shall adopt the Einstein summation convention throughout (as explained e.g. in Penrose and Rindler 1984), unless explicitly indicated otherwise.

It will be beneficial for the reader to be familiar with basics of electromagnetic scattering, diffusion theory, some basic measure theory, and the Ito stochastic calculus (Oksendal 1998), the latter as far as we have developed in Part I.

5

DIFFUSION MODELS OF SCATTERING

Diffusion models have proved highly successful in their application to a large number of problems in applied physics, as described e.g. in Risken (1989). In this chapter we consider the diffusion dynamics underlying electromagnetic scattering processes from random media, such that the asymptotic distribution for the scattered intensity is the K-distribution (Jakeman 1980; Jakeman and Tough 1988; Tough 1987; Ward 1981).

Our analysis does not entail the usual separation of a signal from noise. On the contrary we should think of the pure signal as an inherently stochastic process, a feature which makes our study of particular interest. Accordingly, our purpose is to characterize the noise signal by identifying certain parameters in stochastic differential equations (SDEs) that are used to model the process, namely, the drift and volatility (diffusion) coefficients. In particular, the stochastic volatility coefficient is of special significance since it is independent of the probability measure used to define the underlying Wiener process of the SDE, and is therefore instantaneously observable without recourse to an ensemble average. Identification of the volatility function enables a comparison to be made between the theoretically predicted and empirically observed volatilities, as discussed in terms of real experimental data in Part III. In the K-distributed domain these volatilities should correlate very strongly, and thus one is led to a method for anomaly detection based on where such correlation becomes weak.

The chapter is organized as follows. In Section 5.2 we introduce the electromagnetic scattering process as a vector diffusion process in the cross-section and the scattered intensity, under the assumption that the scattered intensity and intensity scattered from single scatterer are the K-distribution and Rayleigh distribution, respectively. A (partial) classification is thus made of vector diffusion processes such that these marginal distributions are preserved. The volatility coefficients for the component processes emerge as free functions subject to certain natural boundary requirements which are derived.[11] Given the choice of volatility function, preservation of the joint distribution implies that the drifts are accordingly constrained to have a precise form. We complete such classification in Section 5.3 via a detailed investigation of the correlation properties amongst the components of such a vector scattering diffusion model, in terms of their relationship to the asymptotic distributional properties and the conditions for stochastic equilibrium and detailed balance.

---

[11]The situation as regards natural boundaries should be compared with the account of the Bessel process given in Section 4.2.

We observe that from a physical point of view that only the volatility functions are observable and explain this feature in the context of Girsanov's theorem which effects a change of probability measure on the space of all possible paths of the process. We are thus led to an infinite dimensional family of stochastic volatility models for the scattering process such that the discriminating features depend only on the instantaneous observed volatilities and are thus drift independent. In this connection, we remark that similar techniques involving change of measure and stochastic volatility have been studied in the context of financial mathematics, in particular, in the Black–Scholes theory of option pricing (see Appendix D).

5.1 Non-Gaussian statistical models

Earlier work has shown that the distribution of the electromagnetic intensity z for scattering from an object with a fixed (δ-function distribution) scattering cross-section x, equal to the expected value of z, is given by the negative exponential distribution (or equivalently the *Rayleigh distribution* for the square root intensity)

$$\mathcal{R}_{\langle z \rangle}(z) = \frac{1}{\langle z \rangle} \exp\left(\frac{-z}{\langle z \rangle}\right) \tag{5.1}$$

(Jakeman and Tough 1988). The usual treatment of the K-distributed intensity process then makes an assumption that the scattering cross-section x is governed by a birth–death–migration (BDI) population process such that it is asymptotically Γ-distributed and, thereby, one obtains the joint distribution $\mathcal{P}(x, z)$. Integrating out the x variable leads to the *K-distribution* $\mathcal{K}_\nu(z)$ for the (unconditioned) intensity

$$\mathcal{K}_\nu(z) = \frac{2z^{\nu/2} K_\nu(2\sqrt{z})}{\Gamma(\nu + 1)} \tag{5.2}$$

where K_ν is a modified Bessel function of the second kind (see e.g. Jeffreys and Jeffreys 1966).

However, the physical scattering observations relate directly to the scattered radiation, giving rise to the complex-valued amplitude process and corresponding electromagnetic intensity. On the other hand, no direct measurements of the scattering cross-section are made. Thus from a physical point of view the behaviour of the scattering cross-section should be inferred rather than postulated in advance. From large time averages of measurements of the scattered radiation, we can deduce that the intensity is, to a close approximation, K-distributed. The additional assumption of a Rayleigh distributed intensity given a δ-function distributed cross-section (population size) is a direct consequence of a uniform distribution in the phases φ_n of the contributions to the scattered electric field amongst a population of independent identically distributed (i.i.d.) scatterer amplitudes (this conclusion follows readily from the central limit theorem applied to a component of the scattered electric field). More precisely if we decompose the total scattered electric field into the sum of contributions from individual scatterers, thus

$$\mathcal{E} = \sum_n a_n \exp(i\varphi_n), \tag{5.3}$$

then the form factors a_n are taken to be i.i.d. random variables (with any distribution, not necessarily Gaussian) and the phases to be distributed uniformly (cf. Jakeman and Tough 1988). These two features of the electromagnetic scattering, which are physically motivated, then imply the Γ distribution for the scattering cross-section, as seen from the following result.

Lemma 5.1 *For an intensity with $\mathcal{P}(z|x)$ and $\mathcal{P}(z)$ given respectively by the Rayleigh distribution $\mathcal{R}_x(z)$ and the K-distribution $\mathcal{K}_\nu(z)$, the scattering cross-section necessarily has a gamma distribution $\Gamma_\nu(x)$. Accordingly the joint distribution $\mathcal{P}(x, z)$ is equal to $\mathcal{R}_x(z)\Gamma_\nu(x)$.*

Proof Suppose, for some unknown distribution in the cross-section $\mathcal{P}(x)$, that $\mathcal{K}_\nu(z) = \int \mathcal{R}_x(z)\mathcal{P}(x)\mathrm{d}x$, i.e. $\mathcal{K}_\nu(z) = \int [\exp(-z/x)/x]\mathcal{P}(x)\mathrm{d}x$ for some value of ν. Since this $\mathcal{K}_\nu$ is known to be generated by Γ_ν, we have

$$\int \exp\left(\frac{-z}{x}\right)\left(\frac{q(x)}{x}\right)\mathrm{d}x \equiv 0, \quad \forall z, \tag{5.4}$$

where $q = \mathcal{P} - \Gamma_\nu$, so that existence of the above integral is ensured (observe also that $\int q = 0$). Setting $u = 1/x$ the above integral becomes $\int \exp(-uz)\mathcal{Q}(u)\mathrm{d}u$, where $\mathcal{Q}(u) = -q(1/u)/u$, and this integral must vanish for all values of z. Regarding z as a Laplace transform variable, we see from the Laplace transform inversion theorem that $\mathcal{Q} \equiv 0$. Thus $\mathcal{P} = \Gamma_\nu$. $\square$

Observe that the same principle applies to different distributions for the intensity, e.g. if the modulus amplitude is governed by the Weibull distribution, which is sometimes used in radar clutter modelling. More precisely, if the intensity has a given (observed) distribution, and the Rayleigh assumption for the conditional distribution $\mathcal{P}(z|x)$ is preserved, then a corresponding argument leads to the conclusion that the distribution of the cross-section is uniquely determined.

In summary therefore, as a consequence of the various physical assumptions and an observed K-distribution for the intensity, the joint distribution for x, z is given by

$$\mathcal{P}_\infty(x, z) = \frac{x^{\nu-1}\exp(-x - z/x)}{\Gamma(\nu+1)} \tag{5.5}$$

in which $\nu \geq -1$ is the so-called 'shape parameter' as discussed in the radar context in Chapter 13 (equal to $\alpha - 1$ of Chapter 7). We can now proceed to classify (vector) diffusion processes in the variables x, z that possess the above joint distribution. To enable the classification we first observe the following elementary result (cf. also Wong 1963 for a discussion of sufficient conditions for a diffusion to have certain asymptotic distributions, governed by Pearson's equation).

Lemma 5.2 *Given the Fokker–Planck equation (FPE) $\partial \mathcal{P}/\partial t = -\partial_x(\beta \mathcal{P}) + \partial_x^2(\Sigma \mathcal{P})$ the relations between the drift, volatility and asymptotic distribution $\mathcal{P}_\infty$, if such exists, can be summarized as*

$$\mathcal{P}_\infty = \frac{K}{\Sigma} \exp\left(\int^x \frac{\beta}{\Sigma}\right) \tag{5.6}$$

$$\beta = \Sigma \partial_x \log(\Sigma \mathcal{P}) \tag{5.7}$$

$$\Sigma = \frac{k}{\mathcal{P}} + \frac{\int^x \beta \mathcal{P}}{\mathcal{P}}. \tag{5.8}$$

Thus any two of $\{\mathcal{P}, \beta, \Sigma\}$ implies the other (modulo k in the case of Σ).[12]

Proof The existence of an asymptotic distribution $\mathcal{P}_\infty$ implies that $\partial_x[-\beta \mathcal{P}_\infty + \partial_x(\Sigma \mathcal{P}_\infty)] \equiv 0$. Thus the square-bracketed expression is some function $f(t)$, which must vanish given the decay of $\mathcal{P}_\infty(x)$ as $x \to \infty$. Therefore $\beta \mathcal{P}_\infty = \partial_x(\Sigma \mathcal{P}_\infty)$, whereupon the required expression for β is immediate, and integration yields those for $\mathcal{P}_\infty$, Σ. □

5.2 Dynamics of the vector scattering process

This section provides a description of the stochastic dynamics of the processes for the scattering cross-section and scattered intensity, a vector scattering process (x_t, z_t), under the assumption that the intensity for a point scatterer is Rayleigh distributed and the observed scattered intensity is K-distributed (Ward 1981; Jakeman and Tough 1988). The point of view, at this stage, is to generate a large class of dynamics that have appropriate statistical properties. This construction is non-unique, but identifies a certain parametric sub-class of models – thus it is appropriate to situations where one desires to simulate data that is known to possess certain statistical characteristics but, perhaps, where one is less concerned with having a detailed physical understanding of the origins of these dynamics. A more principled and constructive approach, that isolates a specific dynamics uniquely from this class, can be taken essentially from first principles. Such an approach, in contrast, derives statistical properties as emergent consequences of the detailed physical model, and will be provided to enhance and complement the current treatment later in Chapter 8.

We begin by exploiting the results of Lemma 5.2 from which we are able to deduce, for $\mathcal{P}_\infty(x, z)$ prescribed by (5.5), a general form of the FPE for the time-dependent *joint* distribution $\mathcal{P}(x, z, t)$. For the moment, for simplicity, and also to make contact with the work of previous authors (Jakeman and Tough 1988; Tough 1987), we shall neglect the correlation between x and z – later in Section 5.3 we shall develop the classification theory in a more general setting that includes such correlation which in turn corresponds to the construction of K-scattering dynamics provided in Chapter 8.

---

[12] The quantities k and K are constants of integration, equal to each other if the lower limits of integration in the above system are chosen to coincide.

SCATTERING DYNAMICS

Proposition 5.3 *The FPE for K-scattering processes takes the parametric form*

$$\frac{\partial \mathcal{P}}{\partial t} = \mathcal{A}\left[\frac{\partial^2}{\partial x^2}(\phi\mathcal{P}) + \frac{\partial}{\partial x}\left(\left(\left\{\frac{1-\nu}{x} - \frac{z}{x^2} + 1\right\}\phi - \frac{\partial \phi}{\partial x}\right)\mathcal{P}\right)\right]$$
$$+ \mathcal{B}\left[\frac{\partial^2}{\partial z^2}(\psi\mathcal{P}) + \frac{\partial}{\partial z}\left(\left(\frac{\psi}{x} - \frac{\partial \psi}{\partial z}\right)\mathcal{P}\right)\right] \qquad (5.9)$$

for squared volatility functions $\phi(x,z,t)$, $\psi(x,z,t)$. The corresponding non-linearly coupled SDEs for the vector diffusion process (x_t, z_t) are

$$dx_t = \mathcal{A}\left[\left(\frac{\nu-1}{x} + \frac{z}{x^2} - 1\right)\phi + \frac{\partial \phi}{\partial x}\right]dt + (2\mathcal{A}\phi)^{1/2}dW_t^{(1)} \qquad (5.10)$$

$$dz_t = \mathcal{B}\left[\frac{\partial \psi}{\partial z} - \frac{\psi}{x}\right]dt + (2\mathcal{B}\psi)^{1/2}dW_t^{(2)}, \qquad (5.11)$$

where $W_t^{(1)}, W_t^{(2)}$ are independent Wiener processes.

Proof For equilibrium, set the left-hand side of (5.9) equal to zero. Since $\mathcal{A}, \mathcal{B}$ are independent constants we require that the expressions in square brackets on the right-hand side vanish separately. The form of the joint distribution (5.5) together with (5.7) imply the forms taken by the drift coefficients in (5.9), given the squared volatility free functions ϕ, ψ. □

The reciprocals of the (characteristic frequency) constants $\mathcal{A}, \mathcal{B}$ represent independent correlation timescales for the respective scattering processes x_t, z_t. In tensor notation (Risken 1989) we introduce the operator $\hat{L} = -\partial_i \beta^i + \frac{1}{2}\partial_i \partial_j \sigma^{ij}$ in terms of which the FPE can be expressed as $\partial \rho/\partial t = \hat{L}\rho$. The *detailed balance condition*, which is stronger than the stationarity (i.e. equilibrium) condition, requires that each individual transition in the master equation is perfectly balanced. This condition can be expressed as the operator equation $\hat{L}\rho = \rho \hat{L}^\dagger$ acting on all functions f (Risken 1989), whereby the choice $f = 1$ implies stationarity. Alternatively, in terms of the 'current' $v^i = \beta^i - \frac{1}{2}\rho^{-1}\partial_j(\sigma^{ij}\rho)$ the FPE can be written $\partial \rho/\partial t + \partial_i(\rho v^i) = 0$ and the detailed balance condition becomes $v^i = 0$. Since the constants $\mathcal{A}, \mathcal{B}$ in (5.9) are independent and arbitrary, the detailed balance condition is satisfied for any choice of volatility functions ϕ, ψ. In typical physical situations, such as those we shall describe in Chapter 12, z_t de-correlates much more rapidly than its companion x_t, so that $\mathcal{A} \ll \mathcal{B}$. The frequency $\mathcal{A}$ is inherent to the scattering population and is independent of the incident electromagnetic field; in contrast, on dimensional grounds $\mathcal{B}$ is proportional to the wave-number of the illuminating radiation and thus $\mathcal{B} \sim c|\mathbf{k}|$.[13] The (reciprocals of the) constants $\mathcal{A}, \mathcal{B}$ can be determined experimentally from the scattered electromagnetic intensity time series $\{z_t\}$ by measuring the period between successive peaks over the two characteristic long and short time scales,

---

[13] The $\sim$ symbol here indicates an (approximate) scaling relation.

respectively (see e.g. §6 in Jakeman and Tough 1988); cf. also §IIC in Field and Tough (2003b) for a detailed theoretical account of the observability of these characteristic frequency constants.

5.2.1 Natural boundaries

The squared volatility coefficients $\phi(x,z,t)$, $\psi(x,z,t)$ are *free functions* subject to the natural boundary condition, as follows. Since the physical quantities x,z are positive, we require the set $\{x = 0\} \cup \{z = 0\}$ for $x, z > 0$ to bound the vector process (x_t, z_t). Thus the volatilities of the component processes must vanish at the origin and the drifts must tend to positive (possibly infinite) values. Such a requirement ensures that (x_t, z_t) remains in the positive quadrant of the (x, z) plane without imposing additional boundary conditions. Thus the positivity property becomes an inherent feature of the stochastic dynamics and is correspondingly easier to handle. We shall assume that both functions ϕ, ψ are twice differentiable in x, z, respectively, and that these derivatives remain bounded in some closed interval $[0, x_0]$, $x_0 > 0$ (but are not necessarily continuous there). This requirement is natural given the presence of the second derivatives in the FPE for the joint distribution (5.9). Since the natural boundary condition requires $\phi, \psi \to 0$ as $x, z \to 0$ respectively, we may write, for sufficiently small $x > 0$[14]

$$\phi(x) = x\phi^{(1)}(0) + R_\phi(x), \quad (5.12)$$

$$\phi^{(1)}(x) = \phi^{(1)}(0) + R'_\phi(x), \quad (5.13)$$

and likewise for $\psi(z)$ with remainder term $R_\psi(z)$. The Taylor theorem with remainder (see e.g. Jeffreys and Jeffreys 1966) states that $R_\phi(x) = \int_0^x (x-u) \phi^{(2)}(u) du$ and that this remainder satisfies the inequalities

$$\frac{1}{2}x^2 \inf_{[0,x]} [\phi^{(2)}] \leq R_\phi(x) \leq \frac{1}{2}x^2 \sup_{[0,x]} [\phi^{(2)}] \quad (5.14)$$

$$x \inf_{[0,x]} [\phi^{(2)}] \leq R'_\phi(x) \leq x \sup_{[0,x]} [\phi^{(2)}] \quad (5.15)$$

with corresponding inequalities for $\psi(z)$. Thus we deduce the limiting behaviour of the drifts $b^{(x)}$, $b^{(z)}$ in the respective component SDEs (5.10), (5.11). We find

$$\lim_{x \to 0} [b^{(x)}] = \nu\phi^{(1)}(0) + z \lim_{x \to 0}\left(\frac{R_\phi}{x^2}\right) + z\phi^{(1)}(0) \lim_{x \to 0}\left(\frac{1}{x}\right). \quad (5.16)$$

The second term on the right-hand side is bounded by virtue of (5.14). Thus necessary and sufficient conditions for the natural boundary requirement

---

[14] We suppress the (z, t) and (x, t) arguments in the sets of functions $\{\phi, R_\phi\}$, $\{\psi, R_\psi\}$ respectively for notational clarity.

$0 < \lim_{x\to 0} b^{(x)} \leq \infty$ are $\phi^{(1)} > 0$, or else $\phi^{(1)}(0) = 0$ with $\int_0^x (x-u)\phi^{(2)}(u)du > 0$, and sufficient that $\inf_{[0,x]}[\phi^{(2)}] > 0$. Correspondingly for $b^{(z)}$, we find

$$\lim_{x\to 0}[b^{(z)}] = \psi^{(1)}(0) \tag{5.17}$$

so that a necessary and sufficient condition on the component drift for the natural boundary to exist with respect to z is $0 < \psi^{(1)}(0) \leq \infty$.

As a special case of (5.9) (Jakeman and Tough 1988; Tough 1987) when $\phi(x,z,t) = x$ and $\psi(x,z,t) = z$ we find

$$\frac{\partial \mathcal{P}}{\partial t} = \mathcal{A}\left[\frac{\partial^2}{\partial x^2}(x\mathcal{P}) + \frac{\partial}{\partial x}\left(\left(x - \nu - \frac{z}{x}\right)\mathcal{P}\right)\right]$$
$$+ \mathcal{B}\left[\frac{\partial^2}{\partial z^2}(z\mathcal{P}) + \frac{\partial}{\partial z}\left(\left(\frac{z}{x} - 1\right)\mathcal{P}\right)\right], \tag{5.18}$$

where $\mathcal{P} = \mathcal{P}(x,z,t)$. Observe that the natural boundary condition is satisfied for this choice of volatility functions. The associated non-linearly coupled SDEs (5.10), (5.11) for the vector diffusion process (x_t, z_t) become

$$dx_t = \mathcal{A}\left(\frac{\nu - x + z}{x}\right)dt + (2\mathcal{A}x)^{1/2}dW_t^{(1)} \tag{5.19}$$

$$dz_t = \mathcal{B}\left(\frac{1-z}{x}\right)dt + (2\mathcal{B}z)^{1/2}dW_t^{(2)}. \tag{5.20}$$

This special case can be motivated by the corresponding situation for the Rayleigh (Gaussian speckle) process as described in Section 6.1.

In respect of classification of electromagnetic scattering processes, the novelty of the present approach rests on the following key observation.

Lemma 5.4 *The volatility functions ϕ, ψ are instantaneously observable quantities without recourse to an ensemble average.*

This result, for *infinitesimal* dt, follows from the squares of the Ito differentials (5.10) and (5.11). The implementation of the result for *discrete* sampled (experimental) data is explained later in Chapter 12. According to the Ito manipulation rules $dW_t^2 = dt$, $dW\,dt = dt^2 = 0$, these are

$$dx_t^2 = 2\mathcal{A}\phi\,dt, \tag{5.21}$$
$$dz_t^2 = 2\mathcal{B}\psi\,dt. \tag{5.22}$$

The argument for the local observability of the volatility functions can also be understood from the point of view of Girsanov's theorem (Oksendal 1998 and Appendix D) which states that, via a change of measure on the space of paths, the process $d\tilde{W}_t = dW_t + \gamma_t dt$ can be regarded as a pure Brownian motion. More precisely, if **P**, **Q** are probability measures on the space of paths and if W_t is a **P**-Wiener process, then $\tilde{W}_t$ is a **Q**-Wiener process, where the change of measure process M_t (i.e. the Radon–Nikodym derivative $d\mathbf{Q}/d\mathbf{P}$) is given explicitly by the stochastic integral[15]

$$M_t = \exp\left(-\int_0^t \gamma_s dW_s - \frac{1}{2}\int_0^t \gamma_s^2 ds\right). \tag{5.23}$$

Observe from Ito's formula that $dM_t = -M_t \gamma_t dW_t$ so that M_t is manifestly a martingale with respect to the original measure **P**, i.e. $\mathbf{E}_s^\mathbf{P} M_t = M_s$ for $s \leq t$. In respect of our electromagnetic scattering processes we may re-express (5.10) and (5.11) in the $\mathbf{Q}^{(i)}$ 'drift-neutral' measures as

$$dx_t = (2\mathcal{A}\phi)^{1/2} d\tilde{W}_t^{(1)}, \tag{5.24}$$

$$dz_t = (2\mathcal{B}\psi)^{1/2} d\tilde{W}_t^{(2)}. \tag{5.25}$$

Remarks. *On the instantaneous observability of volatility.* It is of considerable interest how our methodology resembles the Black–Scholes theory of *option pricing* (see e.g. Ch. 12 in Oksendal 1998) in financial mathematics – via the change of measure argument, effected via Girsanov's theorem, the stochastic volatilities are the only observable parameters of the SDEs for an individual sample path of the stochastic process. In accordance with this observation, in the financial context, via the introduction of a 'risk-neutral' measure, the Black–Scholes no-arbitrage option price is independent of the drift in the underlying stock price. In our physical application, the observability argument leads to a criterion for *instantaneous* anomaly detection for which the drifts in the underlying electromagnetic scattering processes are of no relevance.

The significance of the volatility over the drift in respect of observability represents a fundamental shift of emphasis in the approach to the problem of anomaly detection in the context of electromagnetic scattering. Previous approaches have used temporal correlations for the intensity process, i.e. quantities such as

$$\langle z_{t_1}^{\alpha_1} z_{t_2}^{\alpha_2} \ldots z_{t_n}^{\alpha_n} \rangle \tag{5.26}$$

which is a moment of order $\alpha_1 + \cdots + \alpha_n$. These moments, while they may provide discriminating measures of data properties, have the shortcoming that they depend on the probability measure **P** and the corresponding diffusion drift. In contrast, the volatility analysis is independent of **P** and requires no ensemble average $\langle \cdot \rangle$.

---

[15] We require γ_t to satisfy Novikov's condition, namely that $\langle \exp(\frac{1}{2}\int_0^t \gamma_s^2 ds)\rangle < \infty$.

5.3 Correlation in the vector scattering process

In the foregoing treatment we have effectively neglected the possibility of correlation between the noise processes $W_t^{(x)}$ and $W_t^{(z)}$. This assumption is represented mathematically by the diffusion tensor *diagonal* (x, z) matrix representation

$$\Sigma^{(\cdot,\cdot)} = \begin{pmatrix} 2\mathcal{A}\phi & 0 \\ 0 & 2\mathcal{B}\psi \end{pmatrix}. \tag{5.27}$$

Now, this special case can be extended to include explicit correlation between $W_t^{(x)}$ and $W_t^{(z)}$. We shall represent this mathematically by a non-zero off diagonal term in the diffusion matrix $\Sigma^{(x,z)} = 2\mathcal{C}\chi$ for some correlation function $\chi = \chi(x, z, t)$, where $\mathcal{C}$ is an (independent) constant with the dimensions of frequency whose reciprocal represents the *cross*-correlation timescale.[16] Thus, in the (x, z) representation we shall have

$$\Sigma^{(\cdot,\cdot)} = \begin{pmatrix} 2\mathcal{A}\phi & 2\mathcal{C}\chi \\ 2\mathcal{C}\chi & 2\mathcal{B}\psi \end{pmatrix}. \tag{5.28}$$

In accordance with the theory of the K-distribution, as developed in the context of scattering above, we maintain the requirement that the joint x, z asymptotic distribution be preserved, which we recall is equal to

$$P_\infty(x, z) = \frac{x^{\nu-1} \exp(-x - z/x)}{\Gamma(\nu+1)}. \tag{5.29}$$

We recall also the *joint* FPE

$$\frac{\partial P}{\partial t} = -\partial_i(Pb^i) + \frac{1}{2}\partial_i\partial_j(P\Sigma^{ij}) \tag{5.30}$$

which must have (5.29) as an asymptotic solution, obtained by setting the left-hand side of (5.30) equal to zero.

The joint FPE can alternatively be expressed in terms of the 'current' v^i as

$$\frac{\partial P}{\partial t} + \partial_i(Pv^i) = 0 \tag{5.31}$$

in which $v^i := b^i - \frac{1}{2}P^{-1}\partial_j(P\Sigma^{ij})$. This is the equation of continuity (conservation of probability, or physically mass if P represents the concentration of a cloud of massive particles).

Remarks. *On the relation with fluid dynamics.* The vector v^i is the current familiar from fluid mechanics. Observe that, even in the presence of stochastic behaviour (in the sense of a non-zero Wiener term in the SDE dynamics), the right-hand side of the equation of continuity is zero. Thus, fluid dynamics should *not*

---

[16] In fact, as we shall see in Chapter 8, from the point of view of a random walk model this cross-correlation has a timescale inherited from the (fluctuations in the) scattering cross-section and thus depends on the constant $\mathcal{A}$.

be viewed as an approximation to a stochastic dynamics and associated FPE – rather, it is an exact description for the fluid current v, as *distinct* from the forward drift b, in a stochastic context.

The relationship between the equilibrium and detailed balance conditions in *one* dimension can be established as follows (assuming rapid enough decay in $P(x^i, t)$ at infinity in x). A steady state $(\partial P_\infty/\partial t = 0)$ implies that Pv^i is equal to some function $f(t)$ of time *only*, by virtue of (5.31), whereupon letting $|x| \to \infty$ and using the decay properties of P we must have $f = 0$. It follows that the current vanishes, and thus detailed balance is equivalent to equilibrium. The corresponding situation in higher dimensions is more subtle – we develop this below.

The vanishing of the time derivative of the joint distribution that is a condition for equilibrium implies, via (5.31), that the vector Pv^i is solenoidal, i.e. it has vanishing divergence. In accordance with a basic theorem from vector calculus, it can therefore be expressed in terms of a vector potential $\mathbf{A}$ as $P\mathbf{v} = \nabla \times \mathbf{A}$ which in our case shall be interpreted in terms of the embedding of $\mathbf{v}$ (an inherently two-dimensional vector) into Euclidean 3-space, and the potential $\mathbf{A}$ necessarily exists in the three-dimensional space.

Before engaging in the detailed mathematics, we begin with some informal remarks concerning detailed balance in respect of the (x_t, z_t) vector scattering process. Consider a visual representation of this process, in the x, z positive quadrant, as an abstract fluid *flow* in case $\mathbf{v} \neq 0$ whose trajectories (streamlines) have abstract 'velocity' $\mathbf{v}$ which is prescribed as the current of the stochastic process (x_t, z_t) according to the ideas of Section 3.2. This constitutes a deterministic mapping of points $\tau : Q_t \mapsto Q_{t+\tau}$ where Q_t has coordinates (x, z) for some fixed initial point, and arises through (5.31) from ensemble average properties of the vector scattering process. Such a flow could occur with fluid density constant in time, i.e. equilibrium, and yet with a non-zero fluid current. Now we ask why, intuitively/physically in terms of this abstract fluid, should the (x_t, z_t) process asymptotically satisfy detailed balance $\mathbf{v} = 0$. Now, this non-vanishing flow can be interpreted as a type of non-stationarity, in the precise sense that it yields canonical *inequivalent times* as pertains to a fluid particle immersed in the current flow (in the asymptotic time domain). This flow becomes an emergent feature of the ensemble average properties of the vector scattering process – its vanishing (detailed balance) is to be regarded as a notion of *strict* equilibrium. In what follows, we shall investigate both the situations where detailed balance is imposed, which we shall regard as a physical constraint, and also for completeness explore what happens if we relax this assumption.

So, suppose that we appeal to our physical argument above and assume strict equilibrium, i.e. detailed balance, motivated in part by our intuition and also by the fact, as we shall see later, that detailed balance is a consequence of the random walk scattering model. Then, we must solve the equation $v^i = 0$, i.e. $Pb^i = \frac{1}{2}\partial_j(P\Sigma^{ij})$. Accordingly, we seek a general solution to the coupled system

$$\begin{cases} Pb^{(x)} = \frac{1}{2}\left[\partial_x(P\Sigma^{xx}) + \partial_z(P\Sigma^{xz})\right], \\ Pb^{(z)} = \frac{1}{2}\left[\partial_x(P\Sigma^{zx}) + \partial_z(P\Sigma^{zz})\right]. \end{cases} \quad (5.32)$$

Before solving this system let us recall briefly the 1-dimensional case. Consider that we are prescribed the asymptotic distribution P and we find the drift for an arbitrary volatility function. Then the system above reduces simply to the scalar relation $b = \frac{1}{2}\Sigma\partial_x[\log(\Sigma P)]$ – thus knowledge of P uniquely establishes the relationship between the drift and volatility (diffusion) coefficients (cf. Wong 1963). In the two-dimensional case of the vector scattering process (x, z) that we are concerned with here, the asymptotic (joint) distribution is indeed prescribed according to (5.29) and thus we follow a similar approach to the one-dimensional case. First, let us recall some simple identities for the partial derivatives of the joint distribution with respect to x, z:

$$\begin{cases} \dfrac{\partial P}{\partial x} = P\left(-1 + \dfrac{z}{x^2} + \dfrac{\nu - 1}{x}\right), \\ \dfrac{\partial P}{\partial z} = -\dfrac{P}{x}. \end{cases} \quad (5.33)$$

From (5.32) this leads us to an expression for the x-drift as

$$2b^{(x)} = \left[\partial_x \Sigma^{(x,x)} + \Sigma^{(x,x)}\left(-1 + \frac{z}{x^2} + \frac{\nu - 1}{x}\right) + \partial_z \Sigma^{(x,z)} + \Sigma^{(x,z)}\left(\frac{-1}{x}\right)\right]. \quad (5.34)$$

and similarly for the z-drift

$$2b^{(z)} = \left[\partial_x \Sigma^{(x,z)} + \Sigma^{(z,x)}\left(-1 + \frac{z}{x^2} + \frac{\nu - 1}{x}\right) + \partial_z \Sigma^{(z,z)} + \Sigma^{(z,z)}\left(\frac{-1}{x}\right)\right]. \quad (5.35)$$

Thus, analogous to the one-dimensional case, we have an explicit mapping from $\Sigma^{(\cdot,\cdot)}$ to $b^{(\cdot)}$ provided by the joint distribution P. We remark that, although our interest here is confined to the vector scattering case (5.29), the methods we have developed would apply equally well to any abstract joint distribution with suitable decay properties, and may therefore also be of pure mathematical interest in their own right.

Given this explicit relationship between the drift and volatility functions of the vector scattering process, which incorporates the presence of cross-correlation, we can verify the generalized dynamics for a special K-scattering model, derived from the physical considerations of a random walk model. For this we take the diffusion tensor to have the special form[17]:

$$\Sigma^{ij} = 2 \begin{pmatrix} \mathcal{A}x & \mathcal{A}z \\ \mathcal{A}z & \mathcal{A}z^2/x + \mathcal{B}xz \end{pmatrix}. \tag{5.36}$$

Observe, as remarked above, that the off-diagonal cross-correlation term scales with the constant $\mathcal{A}$, which originates from the fluctuations in the scattering cross-section process. Note however that the cross-correlation function χ coincides with the intensity variable. Substituting the matrix expression for $\Sigma^{(x,z)}$ into (5.34), and upon cancellation of various terms, we obtain for the x-drift

$$b^{(x)} = \mathcal{A}(\nu + 1 - x). \tag{5.37}$$

Similarly for the z-drift a straightforward calculation using (5.35) and arranging terms according to $\mathcal{A}$ and $\mathcal{B}$ dependence, yields

$$b^{(z)} = \mathcal{A}\left[\frac{z(\nu + 1 - x)}{x}\right] + \mathcal{B}(x - z) \tag{5.38}$$

in accordance with the analysis of the random walk model as developed in Chapter 8. This completes our discussion of the vector K-scattering process in the case that detailed balance is imposed.

Now, we turn to consider the case of classifying K-scattering models if there is 'weak' equilibrium, i.e. an asymptotic limiting distribution exists but the is a non-zero current for the (x_t, z_t) as described in terms of the abstract fluid flow above. We consider the density-weighted immersion of the two-dimensional flow determined by the vector field $\mathbf{v}$ into Euclidean 3-space, co-ordinatized by (X, Y, Z),[18] thus

$$\mathbf{v} \hookrightarrow \mathbf{V} = P\check{\mathbf{v}} \tag{5.39}$$

in which we define $\check{\mathbf{v}} := (v_x, v_z, 0)$. The (weak) equilibrium condition can then be expressed (locally) as

$$\mathbf{V} = \nabla \times \mathbf{A} \tag{5.40}$$

and the vector potential $\mathbf{A}$ is necessarily a *three*-dimensional object. From the geometry of the immersion we may assume $\partial/\partial Z \equiv 0$, whereas general the vector potential (which is subject to a gauge freedom, as identified below) has all components non-vanishing. Accordingly we have

$$\begin{cases} V_X = \partial_Y A_Z \\ V_Y = -\partial_X A_Z \end{cases} \tag{5.41}$$

---

[17] *Orig.* eqn (3.2) in Field and Tough (2003b); note, for comparison, set $\alpha - 1$ in the latter reference equal to ν herein.

[18] Upper case is used here to distinguish the Euclidean coordinates from the vector scattering process (x, z).

and since V_Z vanishes

$$\partial_X A_Y = \partial_Y A_X. \tag{5.42}$$

Thus, without loss of generality, we may set $\mathbf{A} = (0, 0, A)$ for some scalar potential function $A \equiv A(X, Y)$ and hence $\mathbf{V} = (\partial_Y, -\partial_X, 0)A$. Observe here that A has the status of an arbitrary scalar function (with suitable decay at 'spatial' infinity). This enables us to construct a general expression for the forward drift of the process as

$$P\mathbf{b} = P\mathbf{b}_{\text{D.B.}} + (\partial_Y, -\partial_X, 0)A \tag{5.43}$$

consisting on the right-hand side of a detailed balance part and current contribution respectively, wherein the former is obtained by setting $A = 0$. In respect of the latter, we identify the current in terms of a gauge freedom in choice of this scalar function. Observe that the relation $P\mathbf{b}_{\text{D.B.}} = \frac{1}{2}\partial.(\Sigma^{..}P)$ is unaffected by the presence of a gauge term involving A. We therefore summarize the scheme for classifying the dynamics of asymptotic equilibrium as follows: we begin with a prescribed asymptotic joint distribution (e.g. that appropriate to K-scattering), choose the diffusion coefficients $\Sigma^{(.,.)}$ and gauge potential A as free (independent) functions, and then construct the drift according to (5.43) – this provides a complete classification of the diffusion dynamics, incorporating both detailed balance and current contributions. Functionally, we can summarize this situation by the statement

$$\mathbf{b} = \mathbf{b}_{\text{D.B.}}[P, \Sigma] + P^{-1}(\partial_Y, -\partial_X, 0)A. \tag{5.44}$$

In Chapter 8, we shall then see how *specific* choice of the functions ϕ, ψ, χ occurring in the (x, z) representation of $\Sigma^{(.,.)}$ may be obtained, from first principles, via considerations of a random (phasor) walk model. It emerges that detailed balance is satisfied for such a model and so we may assume $A = 0$ in (5.44). Thus, a physical – as opposed to parametric data driven – model emerges as a special case of the general structure above. The latter is constructed merely to preserve the joint (x, z) asymptotic distribution, while the former is a specific instance of this with the essential property that, in addition, it specifies all correlation information and higher order statistics.

6

RAYLEIGH SCATTERING

In this chapter we develop the stochastic dynamical theory of Rayleigh scattering, from the point of view of a random walk model. Physically this is appropriate to forward and backward scattering experiments where the scattering population is effectively constant over the observation period. From a theoretical point of view, the purpose of the chapter is to present a self-contained derivation that follows essentially from first principles. We shall see how the development, according to a random walk model pertaining to the components of the scattered field, leads to an elegant mathematical derivation of the Ornstein–Uhlenbeck process for the resultant amplitude. This derivation of the scattering dynamics, motivated by a physical model (the coherent superposition of a number of randomly phased scattered wave components) is of inherent pure mathematical interest, as well as being satisfying from a theoretical physics point of view.

The chapter is organized as follows. We begin in Section 6.1 with a basic account of the dynamics of the quadrature components of the Rayleigh scattering process, from the point of view of constructing these dynamics from its key statistical properties, such as certain correlation functions and higher order statistics. Given such (limited) statistical characteristics one can consistently posit a set of Ornstein–Uhlenbeck SDEs for the quadrature amplitude components, such that the appropriate statistics emerge. This approach is made physically and mathematically complete, in Section 6.2 by considering the inverse construction, namely that of deriving statistical characteristics from first principled dynamical ones. As opposed to merely constructing a (non-unique) dynamics from desired statistics, we posit a random walk (phase diffusion) model for the scattered amplitude. It is shown, from this point of view, in the Rayleigh case of a fixed step number, that the amplitude *necessarily* obeys a complex Ornstein–Uhlenbeck (COU) equation. A corresponding SDE in the K-scattering case, incorporating step number fluctuations, shall be derived later in Chapter 8.

6.1 Quadrature components

It is familiar that the complex-valued amplitude or 'received envelope' can be expressed as the sum of its *quadrature* components according to

$$\Psi_t = I_t + iQ_t \tag{6.1}$$

so that I, Q denote the 'in-phase' and 'quadrature-phase' components of the coherent amplitude (e.g. radar signal), respectively (e.g. in Helstrom 1960). The intensity z_t can then be expressed in terms of these quantities according to

$$z_t = |\Psi_t|^2 = I_t^2 + Q_t^2. \tag{6.2}$$

Now, in the Rayleigh case, it is consistent from a statistical point of view to describe I_t, Q_t as a pair of *Ornstein–Uhlenbeck* processes (see e.g. Oksendal 1998) with SDEs

$$dI_t = -\alpha I_t dt + \sqrt{2} dW_t^{(I)} \tag{6.3}$$

$$dQ_t = -\alpha Q_t dt + \sqrt{2} dW_t^{(Q)}. \tag{6.4}$$

The Fokker–Planck equation (FPE) associated with either of these separately is

$$\frac{\partial \mathcal{P}}{\partial t} = \frac{\partial^2 \mathcal{P}}{\partial I^2} + \alpha \frac{\partial (I\mathcal{P})}{\partial I} \tag{6.5}$$

whose stationary solution is

$$\mathcal{P}(I) = \left(\frac{\alpha}{2\pi}\right)^{1/2} \exp\left(\frac{-\alpha I^2}{2}\right). \tag{6.6}$$

This shows that I_t is an asymptotically Gaussian-distributed random variable with mean zero and variance $\langle I^2 \rangle = 1/\alpha$ and thus, asymptotically, we have $\langle z \rangle = 2/\alpha$.

6.1.1 Rayleigh intensity

To calculate dz_t we require *Ito's formula* (see e.g. Oksendal 1998 and Chapter 2) which states that, for an Ito process X_t satisfying $dX_t = \mu_t dt + \sigma_t dW_t$ (where μ_t, σ_t are general stochastic processes, not necessarily Ito processes), the process $F_t = f(X_t, t)$ has the stochastic differential $dF_t = \left(\frac{\partial f}{\partial t} + \mu_t \frac{\partial f}{\partial X_t} + \frac{1}{2}\sigma_t^2 \frac{\partial^2 f}{\partial X_t^2}\right) dt + \sigma_t \frac{\partial f}{\partial X_t} dW_t$. Hence the SDE satisfied by z_t is derived according to

$$\begin{aligned} dz_t &= 2I_t dI_t + 2Q_t dQ_t + dI_t^2 + dQ_t^2 \\ &= [4 - 2\alpha(I_t^2 + Q_t^2)]dt + 2\sqrt{2} I_t dW_t^{(I)} + 2\sqrt{2} Q_t dW_t^{(Q)}. \end{aligned} \tag{6.7}$$

Since the sum of two independent Gaussian random variables is itself a Gaussian random variable (see Appendix A), whose variance is the sum of the variances of its constituent parts, we write

$$2\sqrt{2} I_t dW_t^{(I)} + 2\sqrt{2} Q_t dW_t^{(Q)} = 2\sqrt{2z_t} dW_t^{(z)}. \tag{6.8}$$

The SDE for z_t therefore becomes

$$dz_t = (4 - 2\alpha z_t)dt + 2\sqrt{2z_t} dW_t^{(z)} \tag{6.9}$$

whose associated FPE is

$$\frac{\partial \mathcal{P}}{\partial t} = \frac{\partial^2 (4z\mathcal{P})}{\partial z^2} - \frac{\partial ((4 - 2\alpha z)\mathcal{P})}{\partial z}. \tag{6.10}$$

6.1.2 Statistical properties

The stationary solution of this equation is

$$\mathcal{P}(z) = \frac{\alpha}{2} \exp\left(\frac{-\alpha z}{2}\right) \tag{6.11}$$

and thus the mean value of z is what we expect from (6.6). In this way the analysis of the component I, Q processes using the Ornstein–Uhlenbeck process gives insight into the origin of the square root volatility for the intensity z_t. It is interesting to note that the FPE describing the Rayleigh process also encodes the factorization properties of the intensity distribution, which result from its being the sum of squares of two Gaussian processes, each with its own characteristic factorization properties. If we change variables to $z' = \alpha z/2$, $t' = 2\alpha t$, the FPE (6.10) becomes

$$\frac{\partial \mathcal{P}}{\partial t} = \frac{\partial^2 (z\mathcal{P})}{\partial z^2} + \frac{\partial ((z-1)\mathcal{P})}{\partial z}. \tag{6.12}$$

As discussed in detail in Jakeman and Tough (1988), the Green's function of this FPE can be expanded in terms of Laguerre polynomials which, in turn, encode the combinatorial factors implicit in the factorization properties of the Gaussian random variables I_t, Q_t. Thus, for example, we find the exponential decay of the two-time correlation function according to

$$\langle z(t)^n z(0)^m \rangle = n!m! \sum_{r=0}^{\min(m,n)} \frac{n!m!}{(n-r)!(m-r)!(r!)^2} \exp(-rt). \tag{6.13}$$

In much the same way, it can be shown that the expansion of the Green's function of the FPE (6.5) in terms of Hermite polynomials captures the factorization properties of the constituent Ornstein–Uhlenbeck processes. In each case the linearity of the deterministic part of the underlying SDE establishes the exponential decay in the correlation function characteristic of a Markov process.

The constant volatility in the Ornstein–Uhlenbeck case does not impose a natural boundary at the origin, so that the process can take positive and negative values. Conversely, the square root volatility emerging in the Rayleigh case establishes a natural boundary that maintains the positivity of z_t. Observe that the forms taken by the SDEs (6.3) and (6.4) inherently capture these fundamental properties of the physical processes they describe. The special cases (5.18), (5.19) and (5.20) will be significant in the derivation of K-scattering dynamics from a random walk model in Chapter 8 and in the experimental analysis of Chapter 12.

Remarks. *On random motion in a potential.* The situation of stochastic volatility discussed here should be compared with the more familiar case of constant

RANDOM WALK   43

diffusion coefficient and the description of a particle at location x_t in a potential $V(x)$ subject to a random force. The motion is governed by the SDE

$$\mathrm{d}x_t = -\alpha \left(\frac{\partial V}{\partial x}\right) \mathrm{d}t + (2D)^{1/2}\mathrm{d}W_t \tag{6.14}$$

and the particle location has the equilibrium *Boltzmann distribution*

$$\mathcal{P}_\infty(x) \propto \exp\left(\frac{-\alpha V}{D}\right). \tag{6.15}$$

6.2 Random walk model

We develop the random walk model with step number fluctuations due to Jakeman (*orig.* Jakeman 1980; cf. Jakeman and Tough 1988) as a continuous time diffusion process.

In the Rayleigh case consider the random walk model for the scattered electric field (cf. Jakeman 1980; Tough 1987; Jakeman and Tough 1988)

$$\mathcal{E}_t^{(N)} = \sum_{j=1}^{N} \exp\left[i\varphi_t^{(j)}\right] \tag{6.16}$$

for constant population size N. Since Maxwell's equations for the electromagnetic field possess $U(1)$ gauge invariance with respect to duality rotations, i.e. multiplication by $\exp(i\Lambda)$ for constant Λ (cf. Penrose and Rindler 1984), the assumption of independence of $\{\varphi^{(j)}\}$ implies that these phases are uniformly distributed. Accordingly, in (6.16) the phase factors $\{\exp[i\varphi_t^{(j)}]\}$ are independent and uniformly distributed on the unit circle in the complex plane $\mathbf{C}$. Our (phase) diffusion model therefore takes $\{\varphi_t^{(j)}\}$ as a collection of (displaced) Wiener processes evolving on a suitable timescale,

$$\varphi_t^{(j)} = \Delta^{(j)} + \mathcal{B}^{\frac{1}{2}}W_t^{(j)}, \tag{6.17}$$

with the random initializations $\{\Delta^{(j)}\}$ chosen as a set of independent random variables uniformly distributed on the interval $[0, 2\pi)$. The effect of these initializations is to render each component phasor process $\exp[i\varphi_t^{(j)}]$, and thus the resultant amplitude obtained from their coherent addition, stationary (cf. Remarks *On stationarity* below). The component phase dynamics is simply

$$\mathrm{d}\varphi_t^{(j)} = \mathcal{B}^{\frac{1}{2}}\mathrm{d}W_t^{(j)}, \tag{6.18}$$

with squared volatility $\mathrm{d}\varphi_t^{(j)2} = \mathcal{B}\mathrm{d}t$.

6.2.1 Ornstein–Uhlenbeck process

The resultant amplitude dynamics can now be computed explicitly, using the mathematics of the Ito calculus, as developed in Chapter 2. We shall deduce presently that, from the basic assumptions of the random walk model, the dynamics of Rayleigh scattering is necessarily that of the COU process.

From Ito's formula (e.g. Oksendal 1998; Karatzas and Shreve 1988) the Ito differential of (6.16) is

$$d\mathcal{E}_t^{(N)} = \sum_{j=1}^{N} \left(id\varphi_t^{(j)} - \frac{1}{2} d\varphi_t^{(j)2} \right) \exp\left[i\varphi_t^{(j)} \right]. \tag{6.19}$$

The first term $\sum_{j=1}^{N} id\varphi_t^{(j)} \exp\left[i\varphi_t^{(j)}\right]$ on the right-hand side of (6.19) consists of a sum of independent randomly phased Wiener processes, with variance equal to $\mathcal{B}Ndt$, while the second term is independent of the scatterer label j. Thus from (6.19) we can write

$$d\mathcal{E}_t^{(N)} = -\frac{1}{2}\mathcal{B}\mathcal{E}_t^{(N)}dt + (\mathcal{B}N)^{\frac{1}{2}}d\xi_t, \tag{6.20}$$

where ξ_t is a complex Wiener process satisfying $|d\xi_t|^2 = dt$, $d\xi_t^2 = 0$. The process ξ_t is adapted to the filtration $\mathcal{F}^{(\varphi)} = \bigcup_j \mathcal{F}^{(j)}$, where $\mathcal{F}^{(j)}$ is the filtration appropriate to the component scatterer phase $\varphi_t^{(j)}$. The amplitude process Ψ_t is then defined by $\Psi_t = \lim_{N \to \infty} \left[\mathcal{E}_t^{(N)}/\bar{N}^{\frac{1}{2}}\right]$ and satisfies the SDE

$$d\Psi_t = -\frac{1}{2}\mathcal{B}\Psi_t dt + (\mathcal{B}x)^{\frac{1}{2}}d\xi_t, \tag{6.21}$$

where the continuous valued random variable x, the average scattering 'power', arises from an asymptotically large population via $x = \lim_{N \to \infty} [N/\bar{N}]$. Observe that the process exhibits mean reversion (towards the origin), is of constant volatility, and has an asymptotic stable Gaussian distribution.

Remarks. *On geometry of phase wrapping.* The effect of the exponentiation in (6.16) is to 'phase wrap' the process onto the unit circle in the complex plane. Thus, a coordinate discontinuity in the phase process at the boundary of a coordinate interval of length 2π is mapped to a continuous behaviour on the unit circle. The manifold of the phasor process is thus a cylinder, with symmetry axis corresponding to time, on which the sample paths of the phasor process are continuous (with probability 1).

Remarks. *On stationarity.* With regard to stationarity, observe primarily that, given the random uniform initializations (6.17) the distribution of phase remains uniform over the unit circle at all times – the initial distribution is uniform due to the nature of $\Delta^{(j)}$ and this coincides with the asymptotic distribution. The asymptotic uniform property is most readily apparent by considering the Green's function (1.8) phase wrapped onto the unit circle according to

$$f^{\text{wrap}}(\theta) = \sum_{n=-\infty}^{\infty} f(\theta - 2n\pi), \qquad (6.22)$$

where θ is restricted to a coordinate interval $[0, 2\pi)$ and f is the distribution appropriate to the ordinary Wiener process on **R**.[19] For large times the phase wrapped distribution, given the asymptotic behaviour of (1.8), tends to a uniform distribution on the unit circle. Now, stationarity of the phasor *process* follows from the fact that its propagator (a phase wrapped version of that of the Wiener process) is time translation invariant, combined with the Markov property that facilitates factorization of the multi-time joint distribution of the process into two-point propagators. Each component phasor process is therefore (strict sense) stationary. It follows that the resultant amplitude, according to (6.16), also has this property.[20]

Remarks. *On the complex nature of the derivation.* It is quite intriguing, from a mathematical perspective and from the point of view of being driven by considerations of a fundamentally physical nature (namely wavelike interference), that the COU process[21] can be derived from such a random phasor diffusion model, in contrast to which is its application on a comparatively ad hoc basis in other contexts in the literature, such as financial modelling.

---

[19] Equation (6.22) ensures that f^{wrap} is normalized.

[20] These remarks should be contrasted with the case of the pure Wiener process, which is trivially non-stationary (in both the strict and wide senses) due to the time dependence of its single point distribution (1.8) and autocorrelation function.

[21] A corresponding derivation for *real*-valued Ornstein–Uhlenbeck processes (without involving complexification) appears to be unavailable.

7
POPULATION DYNAMICS

The chapter is organized as follows. We begin in Section 7.1 with a discussion of the dynamics of the scattering cross-section and characterize this process in terms of a higher order transition process for a discrete scattering population. This is specialized to the case of the birth–death–immigration (BDI) model in Section 7.2. As the size of the population becomes asymptotically large, in Section 7.3, we obtain a corresponding limiting description in terms of a normalized continuous process, whose probability density obeys the Kramers–Moyal equation. In doing so we elucidate under what circumstances such a continuous process is a diffusion. As a special case we recover the diffusion model applied to scattering by Jakeman (1980) that corresponds to the BDI model, i.e. for the case of first-order transitions of birth, death and immigration.

7.1 Master equations and the Kramers–Moyal expansion

Consider a first-order master equation (Van Kampen 1961; Risken 1989) for the time evolution of the probability $\mathcal{P}_N(t)$ of the event $X_t = N$ for a continuous time integer-valued population process (cf. Fig. 7.1). Denoting the (time dependent) generation and recombination rates by $G_N(t)$, $R_N(t)$ respectively, we find – from conservation of probability – that

$$\frac{d\mathcal{P}_N}{dt} = G_{N-1}\mathcal{P}_{N-1} - (G_N + R_N)\mathcal{P}_N + R_{N+1}\mathcal{P}_{N+1}. \tag{7.1}$$

This is the discrete analogue of the continuous master equation

$$\frac{\partial P(x,t)}{\partial t} = \int [w(x' \to x)P(x',t) - w(x \to x')P(x,t)]dx'. \tag{7.2}$$

Now, from the identity for the Taylor expansion $f(x+l) \equiv \exp(\pm l\partial/\partial x)f(x)$ we deduce that, for the continuous valued counterpart x_t of the discrete N_t with step size l, the above rate equation can be re-expressed as

$$\frac{\partial P(x,t)}{\partial t} = \left[\exp\left(-l\frac{\partial}{\partial x}\right) - 1\right](G(x,t)P(x,t)) + \left[\exp\left(l\frac{\partial}{\partial x}\right) - 1\right](R(x,t)P(x,t)). \tag{7.3}$$

This expression is to be compared with the Kramers–Moyal expansion for the evolution of the probability density

$$\frac{\partial P}{\partial t} = \sum_{n=1}^{\infty}\left(-\frac{\partial}{\partial x}\right)^n [\mathcal{D}^{(n)}\mathcal{P}] \tag{7.4}$$

where the Kramers–Moyal coefficients are defined as the moments

$$\mathcal{D}^n(x,t) = \frac{1}{n!} \lim_{\delta t \to 0} \left\langle \frac{(X_{t+\delta t} - X_t)^n}{\delta t} \right\rangle. \tag{7.5}$$

Observe in relation to (7.4) the *Pawula theorem* (see e.g. Risken 1989), which states that for the probability density $\mathcal{P}$ to remain positive at all times, the expansion (7.4) must either truncate at second order, i.e. reduce to the *Fokker–Planck equation* (FPE), or be of *infinite* order. Combining (7.3) and (7.4) we see that for a first-order transition process, the Kramers–Moyal coefficients are determined by the generation and recombination rates according to

$$\mathcal{D}^{(n)} = \frac{l^n}{n!} \left[G(x,t) + (-1)^n R(x,t) \right]. \tag{7.6}$$

7.2 Birth–death–immigration processes

In the special case of the BDI model, that is of particular interest to modelling random fluctuations in scattering populations, the generation and recombination rates appearing in (7.1) take the forms

$$G = \lambda N + \nu \tag{7.7}$$
$$R = \mu N \tag{7.8}$$

for positive constants λ, μ, ν, the birth, death and immigration rates, respectively. The asymptotic stationary solution to (7.1) for $\lambda < \mu$ is then the *negative binomial distribution*

$$\mathcal{P}_N(\infty) = \binom{N + \alpha - 1}{N} \frac{(\bar{N}/\alpha)^N}{(1 + \bar{N}/\alpha)^{N+\alpha}} \tag{7.9}$$

where $\binom{a}{b}$ is a binomial coefficient, $\alpha = \nu/\lambda$. We deduce that the asymptotic mean, for $\lambda < \mu$, is given by

$$\bar{N} = \frac{\nu}{(\mu - \lambda)}. \tag{7.10}$$

The reader should consult Appendix E for a more detailed discussion of the properties of the BDI model.

The fluctuations in this process are *non-Gaussian* in nature, as evidenced by the super-Poissonian normalized variance of the population size N. More precisely, we define the *Fano factor* (Fano 1947) according to

$$\mathcal{F} = \frac{\text{Var} N}{\langle N \rangle}. \tag{7.11}$$

In the Gaussian case this is equal to unity, whereas for the negative binomial case (7.9) we find

$$\mathcal{F} = \frac{\langle N \rangle}{\alpha} + 1 \tag{7.12}$$

so that $\mathcal{F} > 1$, which is the property that leads to 'bunching' or 'clustering' in the population (Jakeman and Tough 1988).

7.3 Continuum diffusion limit

In this section we discuss the population dynamics of the cross-section appropriate to the modelling of scattering of electromagnetic radiation from certain types of random media that commonly arise. We extend the analysis and present a generalization of the BDI model for a first-order transition process (Bartlett 1966) in the context of the Kramers–Moyal expansion (Kramers 1940; Moyal 1949). In doing so, we shall elucidate the features of the transition rates in the underlying discrete population process that determine whether the associated continuous (normalized) process is a diffusion.

We begin by writing $x_{\text{nor}} = x/\bar{N}$ so that $\bar{x}_{\text{nor}} = 1$. Then, as $\bar{N} \to \infty$ (for an asymptotically large scattering population), x_{nor} becomes a *continuous* valued random variable. The probability density for the re-scaled population process satisfies

$$\mathcal{P}(x,t) \equiv P_N(t) = P_{\bar{N}x}(t). \tag{7.13}$$

Setting the series expansion parameter $l = 1$, the evolution of this density is determined by

$$\frac{\partial \mathcal{P}}{\partial t} = \sum_{n=1}^{\infty} \frac{1}{\bar{N}^n} \left(-\frac{\partial}{\partial x_{\text{nor}}}\right)^n [\mathcal{D}^{(n)}\mathcal{P}] \tag{7.14}$$

or in terms of a re-scaled time parameter $t' = t/\bar{N}$

$$\frac{\partial \mathcal{P}}{\partial t'} = \sum_{n=1}^{\infty} \frac{1}{\bar{N}^{n-1}} \left(-\frac{\partial}{\partial x_{\text{nor}}}\right)^n [\mathcal{D}^{(n)}\mathcal{P}]. \tag{7.15}$$

In the BDI case we evaluate (7.15) in this limit, as follows. For notational clarity we shall drop the prime and 'nor' suffix on t, x, respectively. From (7.6) the summation of (7.15) contains

$$-\left(\frac{\partial}{\partial x}\right)[\nu(1-x)\mathcal{P}] \tag{7.16}$$

for $n = 1$, which equality is exact for finite $\bar{N}$; $n = 2$ yields

$$\frac{1}{2}\left(\frac{\partial}{\partial x}\right)^2 \{[x(\lambda + \mu) + \nu/\bar{N}]\mathcal{P}\} \tag{7.17}$$

for finite $\bar{N}$ (ν remains bounded as $\bar{N} \to \infty$) which, as $\bar{N} \to \infty$, becomes

$$\left(\frac{\partial}{\partial x}\right)^2 (\lambda x \mathcal{P}). \tag{7.18}$$

For $n \geq 3$ the sum contains derivatives of $\mathcal{D}^{(n)}/\bar{N}^{n-1}$ which, for first-order transitions, is equal to

$$\frac{1}{n!}\frac{(G(x,t)+(-1)^n R(x,t))}{\bar{N}^{n-1}} \tag{7.19}$$

for step size $l = 1$. These terms vanish in the limit $\bar{N} \to \infty$ since G, R and, consequently, $\mathcal{D}^n$, are linear in $\bar{N}$. Thus, the normalized process x_{nor} is a diffusion with probability density governed by the FPE

$$\frac{\partial \mathcal{P}}{\partial t} = -\frac{\partial}{\partial x}[\nu(1-x)\mathcal{P}] + \frac{\partial^2}{\partial x^2}(\lambda x \mathcal{P}). \tag{7.20}$$

In this derivation we have used the expression for the asymptotic mean in terms of the population parameters (7.10). The SDE associated with the above FPE can now be read off as

$$\mathrm{d}x_t = \nu(1-x)\mathrm{d}t + (2\lambda x)^{1/2}\,\mathrm{d}W_t. \tag{7.21}$$

The asymptotic distribution for this process is then $\mathcal{P}_\infty(x) \propto x^{\alpha-1}\exp(-\alpha x)$ ($\alpha = \nu/\lambda$), so the re-scaled variate αx has the *gamma distribution*

$$\Gamma_\alpha(x) = \frac{x^{\alpha-1}\exp(-x)}{\Gamma(\alpha)}, \tag{7.22}$$

where $\Gamma(\alpha)$ is the gamma function (see e.g. Jeffreys and Jeffreys 1966). The situation in respect of the limiting $\bar{N}$ behaviour and the population parameters can therefore be summarized as follows. If we let $\nu \to \infty$, keeping $\bar{N}$ fixed, then the discrete Poisson distribution emerges from (7.9). If one now lets $\bar{N} \to \infty$, the associated continuous distribution can be identified as a delta function. The delta function also occurs if the limits are taken in the reverse order, thus $\bar{N} \to \infty$, via $\mu \to \lambda$ yields a gamma distribution, and $\nu \to \infty$, with fixed mean, reduces the gamma distribution to a delta function.

Remark. The case of the pure *Poisson process* is obtained when the birth and death rates both vanish and there is a constant immigration rate.

For a more general first-order master equation, we expect the transition coefficients G, R to have higher order polynomial behaviour. In this respect we have the following result.

Proposition 7.1 *Given a first-order discrete valued transition process N_t, a necessary and sufficient condition for the associated continuous valued normalized random variable x_{nor} to be of diffusion type (with respect to the re-scaled time t') is that G, R are $\mathcal{O}(N^\alpha)$ for $\alpha < 3$ and $G - R \sim \mathcal{O}(N^\beta)$ for $\beta < 2$.*

Proof Sufficiency is immediate from the $\bar{N} \to \infty$ limiting behaviour of $(G(x,t)+(-1)^n R(x,t))/\bar{N}^{n-1}$. Conversely, the zero contribution for $n = 3$ implies the required behaviour for $G - R$ and thus trivially that $G - R \sim \mathcal{O}(N^\beta)$ for $\beta < 3$. Combining this with the relation $G + R \sim \mathcal{O}(N^\beta)$, $\beta < 3$ which follows from the zero contribution for $n = 4$, yields the required behaviour for G, R separately.

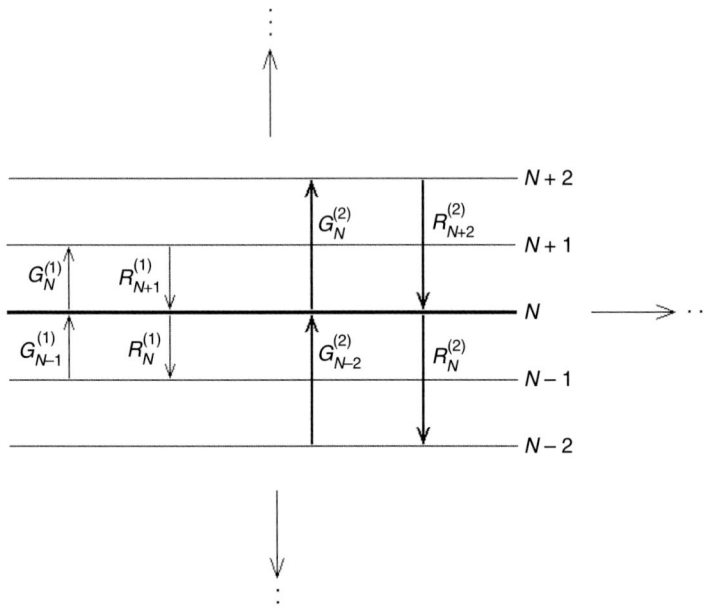

FIG. 7.1. Higher order transition process.

It is of some interest that, even for a first-order master equation, the expansion (7.15) is of infinite order *generically*, and accordingly the fluctuations in the process can not be driven by the usual Wiener process (cf. Jakeman *et al.* 2003). For higher order transitions the situation is depicted in Fig. 7.1. We can express the evolution of the probability density in a similar fashion as

$$\frac{d\mathcal{P}_N}{dt} = \sum_{i=1}^{\infty} \{G^{(i)}_{N-i}\mathcal{P}_{N-i} - (G^{(i)}_N + R^{(i)}_N)\mathcal{P}_N + R^{(i)}_{N+i}\mathcal{P}_{N+i}\}, \quad (7.23)$$

where $G^{(i)}_{N-i} = R^{(i)}_N = 0$ for $i > N$, thus maintaining $N \geq 0$. Correspondingly the Kramers–Moyal coefficients are given by

$$\mathcal{D}^{(n)} = \frac{l^n}{n!} \left[\sum_{i=1}^{\infty} i^n (G^{(i)} + (-1)^n R^{(i)}) \right]. \quad (7.24)$$

The order of the Kramers–Moyal expansion for higher order transition processes can be summarized by the following result.

Proposition 7.2 *The Kramers–Moyal expansion (7.15) is finite order as $\bar{N} \to \infty$ if and only if for all integer $m > 2$*

$$\sum_{i=1}^{\infty} i^m [G^{(i)} + (-1)^m R^{(i)}] = o(N^{m-1}) \quad (7.25)$$

as $N \to \infty$. Correspondingly, the continuous normalized stochastic process x_{nor} is a diffusion with respect to the re-scaled time $t' = t/\bar{N}$.

The alternative possibility is supplied by the Pawula theorem.

We refer the reader to Appendix E for details of the solution to the rate equation for the BDI model via the partition function. A complete treatment of the special limiting cases that arise in that context is also provided therein.

8

DYNAMICS OF K-SCATTERING

We derive the stochastic dynamics of the complex-valued amplitude resulting from coherent scattering from a random population of scatterers when this becomes asymptotically large. Considerations of a random walk model, introduced by Jakeman, are used to derive stochastic differential equations (SDEs) for the amplitude and corresponding intensity and phase stochastic processes. An analysis of the correlation structure in the fluctuations is provided and interpreted geometrically in terms of the gauge invariant properties of the field and the Markov property. A Fokker–Planck description for the evolution of the probability density is given and the equilibrium and detailed balance conditions are shown to hold. Expressions for the intensity autocorrelation function and power spectral density are provided in closed form. The practical implications of the stochastic theory are discussed.

Recent developments in the diffusion-based analysis of scattering from random media, reported in Field and Tough (2003a), have led to significant results that enable the identification of K-distributed noise processes in electromagnetic scattering. The results comprise various SDEs for the scattered amplitude, intensity, phase, and scattering cross-section motivated by a combination of theoretical considerations and analysis of empirical data.

The purpose of the present chapter is to formulate the stochastic dynamics of the electromagnetic field, scattered from a random medium that consists of a collection of independent component scatterers, whose population size has fluctuations in accordance with the birth–death–immigration (BDI) model (Bartlett 1966). This is achieved from first principles via considerations of the complex random walk model introduced by Jakeman (Jakeman 1980). Our results thus provide the theoretical foundation of the anomaly detection technique reported earlier (Field and Tough 2003a), and extend the model to include a detailed description of the intensity autocorrelation function and power spectral density.

The chapter is organized as follows. Section 8.1 recalls the complex amplitude SDE in the case of a constant scattering cross-section, i.e. Rayleigh scattering, that follows from considerations of a complex random walk model, and extends this to the K-distributed case via the insertion of step number fluctuations. The resulting SDE for the amplitude process is used to derive corresponding SDEs for the intensity and phase processes, and expressions for the squared volatilities of each of these processes are provided. This framework is used to explain the geometrical structure of the correlations in the fluctuations of the complex amplitude in Section 8.2.

K-AMPLITUDE

In Section 8.3, we provide a Fokker–Planck description for the joint probability density of the scattering cross-section and intensity processes and study the asymptotic behaviour. It is verified that the model possesses the joint probability appropriate to the K-distribution, and that the condition for detailed balance is also satisfied.

Section 8.4 provides a detailed analysis of the finite-time correlation properties of the stochastic model. The Green's functions arising from the Fokker–Planck description are computed and expressions for the autocorrelation functions and power spectral density are provided in closed form, and the spectral properties discussed. The chapter concludes with an account of the interpretation and experimental implications of the theoretical framework proposed.

8.1 Stochastic dynamics of K-amplitude process

8.1.0.1 K-distributed noise
In the case of step number fluctuations in the random walk model (6.16), we define the amplitude Ψ_t in a similar manner to the Rayleigh case above, with the modification that we employ a time dependent N_t such that $x_t = \lim_{N_t \to \infty} [N_t/\bar{N}]$. Thus

$$\Psi_t = \lim_{N_t \to \infty} \left\{ \frac{1}{\bar{N}^{\frac{1}{2}}} \sum_{j=1}^{N_t} \exp\left[i\varphi_t^{(j)}\right] \right\} \tag{8.1}$$

$$= \lim_{N_t \to \infty} \left\{ \left(\frac{N_t}{\bar{N}}\right)^{\frac{1}{2}} \frac{1}{N_t^{\frac{1}{2}}} \sum_{j=1}^{N_t} \exp\left[i\varphi_t^{(j)}\right] \right\} \tag{8.2}$$

$$= x_t^{\frac{1}{2}} \gamma_t, \tag{8.3}$$

where $\gamma_t = \lim_{N \to \infty} \left[\mathcal{E}_t^{(N_t)}/N_t^{\frac{1}{2}}\right]$.

Remark. *On the failure of the central limit theorem.* It is important to observe the *failure* of central limit theorem to apply in this case, because of the nature of the square root normalization and the fact that the number of steps in the random walk model fluctuates in time.

According to the arguments given in the Rayleigh case above, γ_t is a complex Ornstein–Uhlenbeck (COU) process which obeys the SDE

$$d\gamma_t = -\frac{1}{2}\mathcal{B}\gamma_t dt + \mathcal{B}^{\frac{1}{2}} d\xi_t. \tag{8.4}$$

Observe from (6.21), therefore, that γ_t is a unit power Rayleigh process. The above equation for γ_t can be solved by considering the stochastic differential $d\left[\exp(\frac{1}{2}\mathcal{B}t)\gamma_t\right]$, which leads to the solution

$$\gamma_t = \exp\left(-\frac{1}{2}\mathcal{B}t\right) \left\{ \gamma_0 + \mathcal{B}^{\frac{1}{2}} \int_0^t \exp\left(\frac{1}{2}\mathcal{B}s\right) d\xi_s \right\}. \tag{8.5}$$

We deduce the expectation formulae

$$\mathbf{E}\left[\gamma_t\right] = \exp\left(-\frac{1}{2}\mathcal{B}t\right)\gamma_0, \tag{8.6}$$

$$\mathbf{E}\left[|\gamma_t|^2\right] = 1 + \exp(-\mathcal{B}t)(|\gamma_0|^2 - 1). \tag{8.7}$$

From (8.7) it follows that $\lim_{t\to\infty}\mathbf{E}\left[|\gamma_t|^2\right] = 1$ and so from (8.3) we find the intensity process, defined by $z_t = |\Psi_t|^2$, satisfies

$$\mathbf{E}[z_t | \mathcal{F}_t^{(x)}] = x_t \tag{8.8}$$

where the expectation is conditional on a sharp value of the cross-section and thus averages of the phases $\varphi^{(j)}$ in (6.16) only. The SDE for Ψ_t, as determined by (8.3), can now be derived, via the Ito product formula $\mathrm{d}(X_tY_t) \equiv X_t\mathrm{d}Y_t + Y_t\mathrm{d}X_t + \mathrm{d}X_t\mathrm{d}Y_t$. For this purpose, it is convenient to introduce the square-root cross-section $r_t = x_t^{1/2}$ and thus, from (8.3), we find $\mathrm{d}\Psi_t = r_t\mathrm{d}\gamma_t + \gamma_t\mathrm{d}r_t$. Observe that the cross term $\mathrm{d}r_t\mathrm{d}\gamma_t$ does not feature in this relation owing to the independence of $W_t^{(x)}$ and $\xi_t^{(\varphi)}$, which originate from the intrinsic scattering population and the scattered electromagnetic field, respectively.

The BDI model (Bartlett 1966) posits a first-order master equation for the population N_t, with respective generation and recombination rates $G = \lambda N + \nu$, $R = \mu N$. For an asymptotically large population, $N \to \infty$, we deduce (Tough 1987; Jakeman and Tough 1988; Field and Tough 2003a) that the re-scaled population variate $x \mapsto \alpha x$ satisfies the SDE

$$\mathrm{d}x_t = \mathcal{A}(\alpha - x_t)\mathrm{d}t + (2\mathcal{A}x_t)^{\frac{1}{2}}\mathrm{d}W_t^{(x)} \tag{8.9}$$

for an independent Wiener process $W_t^{(x)}$, where $\alpha = \nu/\lambda$. This leads to the asymptotic Γ-distribution for x_t,

$$\Gamma_\alpha(x) = \frac{x^{\alpha-1}\exp(-x)}{\Gamma(\alpha)} \tag{8.10}$$

so that $\mathrm{Var}[x] = \langle x \rangle = \alpha$. From Ito's formula applied to $r_t = x_t^{1/2}$ we find $\mathrm{d}r_t = \mathrm{d}x_t/2x_t^{1/2} - \mathrm{d}x_t^2/8x_t^{3/2}$, and thus from (8.9)

$$\mathrm{d}r_t = \mathcal{A}\left(\frac{2(\alpha - x_t) - 1}{4r_t}\right)\mathrm{d}t + \left(\frac{\mathcal{A}}{2}\right)^{\frac{1}{2}}\mathrm{d}W_t^{(x)}. \tag{8.11}$$

This leads to the following result.

Theorem 8.1 *(Field equations.) In the K-distributed case the scattered amplitude is governed by the SDE*

$$\frac{\mathrm{d}\Psi_t}{\Psi_t} = -\frac{1}{2}\mathcal{B}\mathrm{d}t + \frac{\mathcal{B}^{\frac{1}{2}}}{\gamma_t}\mathrm{d}\xi_t + \mathcal{A}\left(\frac{2(\alpha - x_t) - 1}{4x_t}\right)\mathrm{d}t + \left(\frac{\mathcal{A}}{2x_t}\right)^{\frac{1}{2}}\mathrm{d}W_t^{(x)}. \tag{8.12}$$

This evolution is invariant under the $U(1)$ gauge transformation $\Psi_t \mapsto \exp(i\Lambda)\Psi_t$, for constant Λ.

Remark. In many respects this result, which is derived from very fundamental physical and mathematical principles, represents the central theoretical advance within the entire body of the monograph.

In the above expression, $\mathcal{A}$ and $\mathcal{B}$ are independent constants with the dimension of frequency, and they may take arbitrary values. In most situations of interest, however, such as those reported in Field and Tough (2003a) and developed in Chapter 12, the wavelength of the illuminating radiation is such that the two corresponding reciprocal correlation timescales satisfy $\mathcal{A} \ll \mathcal{B}$. The description of Rayleigh scattering (i.e. constant scattering cross-section) is then recovered when $\mathcal{A} = 0$. Theorem 8.1 implies the following result.

Corollary 8.2 *The squared volatility of the amplitude process Ψ_t is given by*

$$|\mathrm{d}\Psi_t|^2 = \left(\mathcal{B}x_t + \frac{\mathcal{A}z_t}{2x_t}\right)\mathrm{d}t. \tag{8.13}$$

It is the *linearity* of the right-hand side above in z_t that, in part, enables the anomaly detection mechanism described in Section 12.2 (*orig.* Field and Tough (2003a)).

8.1.1 *Intensity*

The stochastic differential of the intensity process z_t can be expressed in terms of the amplitude via the identity

$$\mathrm{d}z_t = \Psi_t^* \mathrm{d}\Psi_t + \Psi_t \mathrm{d}\Psi_t^* + |\mathrm{d}\Psi_t|^2 \tag{8.14}$$

which follows from $z_t = |\Psi_t|^2$. From (8.4) we find

$$\Psi_t^* \mathrm{d}\Psi_t + \Psi_t \mathrm{d}\Psi_t^* = x_t(\gamma_t^* \mathrm{d}\gamma_t + \gamma_t \mathrm{d}\gamma_t^*) + 2|\gamma_t|^2 r_t \mathrm{d}r_t$$

$$= -\mathcal{B}z_t \mathrm{d}t + \mathcal{B}^{\frac{1}{2}} x_t(\gamma_t^* \mathrm{d}\xi_t + \gamma_t \mathrm{d}\xi_t^*) + \frac{2z_t}{r_t}\mathrm{d}r_t. \tag{8.15}$$

The terms involving $\mathrm{d}\xi_t$ above can be combined in terms of a real-valued Wiener process $W_t^{(\varphi)}$ according to

$$\gamma_t^* \mathrm{d}\xi_t + \gamma_t \mathrm{d}\xi_t^* \equiv \left(\frac{2z_t}{x_t}\right)^{\frac{1}{2}} \mathrm{d}W_t^{(\varphi)}. \tag{8.16}$$

We deduce from (8.13), (8.14), (8.15) that

$$\mathrm{d}z_t = -\mathcal{B}z_t \mathrm{d}t + (2\mathcal{B}z_t x_t)^{\frac{1}{2}}\mathrm{d}W_t^{(\varphi)} + \frac{2z_t}{r_t}\mathrm{d}r_t + \left(\mathcal{B}x_t + \frac{\mathcal{A}z_t}{2x_t}\right)\mathrm{d}t. \tag{8.17}$$

In combination with (8.11) this leads to the following result.

Proposition 8.3 *The intensity SDE is given by*

$$dz_t = \left[\mathcal{B}(x_t - z_t) + \frac{\mathcal{A}z_t(\alpha - x_t)}{x_t}\right]dt + \left(2\mathcal{B}x_t z_t + \frac{2\mathcal{A}z_t^2}{x_t}\right)^{\frac{1}{2}} dW_t^{(z)} \quad (8.18)$$

in which $W_t^{(z)}$ *is correlated with* $W_t^{(x)}$ *of (8.9), and satisfies*

$$\left(\mathcal{B}x_t z_t + \frac{\mathcal{A}z_t^2}{x_t}\right)^{\frac{1}{2}} dW_t^{(z)} = (\mathcal{B}x_t z_t)^{\frac{1}{2}} dW_t^{(\varphi)} + \left(\frac{\mathcal{A}}{x_t}\right)^{\frac{1}{2}} z_t dW_t^{(x)}. \quad (8.19)$$

The filtration of $W_t^{(\varphi)}$ arises from the constituent phases $\varphi_t^{(j)}$ in the random walk according to (8.16), while that of $W_t^{(x)}$ stems solely from the fluctuations in the endogenously specified population model – such (scattering) model is thus sometimes referred to as the *endogenous model*. Observe that, if $\mathcal{B} = 0$, (8.19) implies $W^{(z)} = W^{(x)}$, and from (8.4) γ_t is constant, so $|\gamma_t|^2 = \mathbf{E}\left[|\gamma_t|^2\right] = 1$. Accordingly, $z_t = x_t$ is a solution of (8.18), as required by (8.3). From Proposition 8.3 we obtain the following result.

Corollary 8.4 *The squared intensity volatility is determined by*

$$dz_t^2 = \left(2\mathcal{B}x_t z_t + \frac{2\mathcal{A}z_t^2}{x_t}\right) dt. \quad (8.20)$$

Alternatively, in terms of the amplitude process Ψ_t the squared volatility in the intensity z_t can be expressed as

$$dz_t^2 \equiv \Psi_t^2 d\Psi_t^{*2} + \Psi_t^{*2} d\Psi_t^2 + 2z_t |d\Psi_t|^2 \quad (8.21)$$

which, from (8.12), leads to the above expression for dz_t^2. Observe that for $\mathcal{A} \ll \mathcal{B}$ the dominant contribution to the squared intensity volatility is proportional to the instantaneous value of the intensity. Thus for a Rayleigh timescale $\mathcal{B}^{-1}$, over which x_t remains approximately constant, the time series for dz_t^2 and z_t should exhibit strong correlation. This feature has been experimentally verified in a case of optical scattering, which is described in Section 12.1 (*orig.* §4(*a*) of Field and Tough (2003*a*)).

In terms of the square-root intensity $R_t = \sqrt{z_t}$, an application of Ito's formula to (8.20) yields the following result.

Corollary 8.5 *The squared volatility in the modulus amplitude is determined by*

$$dR_t^2 = \frac{1}{2}\left(\mathcal{B}x_t + \frac{\mathcal{A}z_t}{x_t}\right) dt. \quad (8.22)$$

8.1.2 Phase

The complex amplitude process can be expressed in polar form $\Psi_t = R_t \exp(i\theta_t)$ and thus, writing $i\theta_t = \log(\Psi_t/R_t)$, we deduce from Ito's formula that

$$i d\theta_t = \frac{d\Psi_t}{\Psi_t} - \frac{1}{2}\left(\frac{d\Psi_t}{\Psi_t}\right)^2 - \frac{dR_t}{R_t} + \frac{1}{2}\left(\frac{dR_t}{R_t}\right)^2. \tag{8.23}$$

Since the left-hand side is purely imaginary we can express $d\theta_t$ in terms of Ψ_t alone as

$$d\theta_t = \frac{1}{2i}\left[\left(\frac{d\Psi_t}{\Psi_t} - \frac{1}{2}\left(\frac{d\Psi_t}{\Psi_t}\right)^2\right) - \left(\frac{d\Psi_t^*}{\Psi_t^*} - \frac{1}{2}\left(\frac{d\Psi_t^*}{\Psi_t^*}\right)^2\right)\right]. \tag{8.24}$$

Accordingly, the squared phase volatility is determined by the identity

$$d\theta_t^2 \equiv \frac{|d\Psi_t|^2}{2z_t} - \frac{d\Psi_t^2}{4\Psi_t^2} - \frac{d\Psi_t^{*2}}{4\Psi_t^{*2}}. \tag{8.25}$$

From (8.12) we have

$$\frac{d\Psi_t}{\Psi_t} - \frac{1}{2}\left(\frac{d\Psi_t}{\Psi_t}\right)^2 = \left(\frac{\mathcal{A}(\alpha - x_t - 1)}{2x_t} - \frac{1}{2}\mathcal{B}\right)dt + \left(\frac{\mathcal{A}}{2x_t}\right)^{\frac{1}{2}} dW_t^{(x)} + \frac{\mathcal{B}^{\frac{1}{2}}}{\gamma_t} d\xi_t. \tag{8.26}$$

Hence, from (8.24), θ_t obeys the SDE

$$d\theta_t = \frac{\mathcal{B}^{\frac{1}{2}}}{2i|\gamma_t|^2}(\gamma_t^* d\xi_t - \gamma_t d\xi_t^*). \tag{8.27}$$

As in the derivation of the SDE for the intensity, we can express the terms involving ξ_t in (8.27) as a distinct real-valued Wiener process $W_t^{(\theta)}$ according to

$$\frac{1}{2i}(\gamma_t^* d\xi_t - \gamma_t d\xi_t^*) \equiv \left(\frac{z_t}{2x_t}\right)^{\frac{1}{2}} dW_t^{(\theta)}. \tag{8.28}$$

Thus we obtain the following result.

Proposition 8.6 *The resultant phase θ_t of the complex K-amplitude process Ψ_t obeys the SDE*

$$d\theta_t = \left(\frac{\mathcal{B}x_t}{2z_t}\right)^{\frac{1}{2}} dW_t^{(\theta)} \tag{8.29}$$

which has vanishing drift.

The result has the following consequence.

Corollary 8.7 *The squared resultant phase volatility is given by*

$$d\theta_t^2 = \frac{\mathcal{B}x_t}{2z_t}\,dt. \tag{8.30}$$

Alternatively, (8.30) can be derived by applying (8.25) to (8.12). This result accords with the general scaling and symmetry arguments for the behaviour of the squared phase volatility put forward in §4 of Field and Tough (2003a). The situation should be contrasted with the differentiable model for the process Ψ_t (Jakeman et al. 2001) for which the intensity-weighted phase-derivative, instead of its square, has minimal variance.

These various relations allow the K-amplitude dynamics to be recast in terms of $W^{(x)}$, $W^{(\theta)}$, $W^{(r)}$ as follows.

Corollary 8.8 *The K-amplitude satisfies the SDE*

$$\frac{d\psi_t}{\psi_t} = \left[\mathcal{A}\left(\frac{2(\alpha - x_t) - 1}{4x_t}\right) - \frac{1}{2}\mathcal{B}\right]dt + \left(\frac{\mathcal{A}z_t + \mathcal{B}x_t^2}{2x_t z_t}\right)^{\frac{1}{2}}dW_t^{(z)} + i\left(\frac{\mathcal{B}x_t}{2z_t}\right)^{\frac{1}{2}}dW_t^{(\theta)} \tag{8.31}$$

in which, alternatively, the Wiener terms can be expressed as

$$\left(\frac{\mathcal{A}}{2x_t}\right)^{\frac{1}{2}}dW_t^{(x)} + \left(\frac{\mathcal{B}x_t}{2z_t}\right)^{\frac{1}{2}}\left(dW_t^{(r)} + idW_t^{(\theta)}\right). \tag{8.32}$$

The following result, implied by (8.12) and the identities $d\xi_t^2 = d\xi_t dW_t^{(x)} = 0$, is relevant in connection with the geometry of fluctuations for weak scattering processes discussed in Section 9.3.

Corollary 8.9 *The product cross-section/K-amplitude stochastic differentials satisfy*

$$dx_t^2 = 2\mathcal{A}x_t\,dt, \tag{8.33}$$

$$dx_t d\psi_t = \mathcal{A}\psi_t\,dt, \tag{8.34}$$

$$d\psi_t^2 = \left(\frac{\mathcal{A}\psi_t^2}{2x_t}\right)dt, \tag{8.35}$$

$$|d\psi_t|^2 = \left(\frac{\mathcal{A}z_t}{2x_t} + \mathcal{B}x_t\right)dt. \tag{8.36}$$

The results above lead to expressions for the frequency constants $\mathcal{A}$ and $\mathcal{B}$, as follows. With respect to an average over the phase fluctuations $d\varphi_t^{(j)}$ there exists a *residual constant term* in the squared phase volatility, i.e.

$$\mathbf{E}\left[d\theta_t^2/dt|\mathcal{F}^{(\varphi)}\right] = \frac{1}{2}\mathcal{B}. \tag{8.37}$$

In principle, this enables the Rayleigh constant $\mathcal{B}$ to be deduced from scattering data (alternatively an estimate of the Rayleigh correlation timescale $\mathcal{B}^{-1}$ can be found from measuring the time difference between successive peaks in the intensity time series $\{z_t\}$). Expression (8.30) implies that $x_t = 2\mathcal{B}^{-1} z_t \mathrm{d}\theta_t^2 / \mathrm{d}t$ and so the instantaneous values of the cross-section x_t, and therefore r_t, are observable through the squared phase fluctuations. Consequently, the constant $\mathcal{A}$ can be deduced from the square of (8.11)

$$\frac{\mathrm{d}r_t^2}{\mathrm{d}t} = \frac{1}{2}\mathcal{A} \qquad (8.38)$$

(cf. §3 in Field and Tough (2003a) for an account of the observability of the squared volatilities for discretely sampled time series data).

8.2 Geometry of K-amplitude fluctuations

In relation to the discussion of Chapter 3 and Section 3.1 therein, we apply the concept of diffusion tensor arising in stochastic differential geometry to the detailed amplitude dynamics appropriate to K-scattering. These stochastic dynamics, as presented in Theorem 8.1, enable one to deduce the correlation structure in the fluctuations of the complex amplitude process Ψ_t.

A geometric insight into these properties can be gained from the symmetry properties of the process Ψ_t. Theorem 8.1 shows that the SDE for Ψ_t is invariant under the $U(1)$ gauge transformation $\Psi_t \mapsto \exp(i\Lambda)\Psi_t$. Since the drift in (8.26) is real-valued, the identity (8.24) implies that θ_t has vanishing drift, as seen explicitly from Proposition 8.6. The resulting asymptotic probability distribution is therefore $U(1)$ symmetric.

In respect of time evolution, the SDE (8.12) has the Markov property (see Appendix C), that the evolution it determines depends on the instantaneous value of Ψ_t and is independent of the history of the process $\{\Psi_{t'} | t' < t\}$. This feature yields a preferred symmetry, namely the instantaneous radial direction determined by Ψ_t and its orthogonal θ direction. The diffusion tensor σ^{ij}, determined by $\mathrm{d}\Psi^i \mathrm{d}\Psi^j = \sigma^{ij}\mathrm{d}t$ is real and symmetric and therefore can be diagonalized over $\mathbb{C}$. In the non-degenerate case its eigenvectors constitute a unique orthogonal pair corresponding to the directions in which the component Wiener increments are independent. On grounds of the above symmetry, we anticipate that the space of eigen-directions contains the instantaneous radial and attendant θ direction. This geometrical property can be verified explicitly from (8.16), (8.19), and (8.28), which imply the following.

Proposition 8.10 *The cross-correlation between $W_t^{(z)}$, $W_t^{(\theta)}$ vanishes identically, i.e. $\mathrm{d}W_t^{(z)} \mathrm{d}W_t^{(\theta)} = 0$. Correspondingly $\mathrm{d}R_t \mathrm{d}\theta_t = 0$, i.e. the fluctuations in R_t, θ_t are statistically independent.*

In terms of the I, Q component representation the coordinate transformations $I = R\cos\theta$, $Q = R\sin\theta$ and the property $\mathrm{d}R_t \mathrm{d}\theta_t = 0$ imply the geometric relation

$$\mathrm{d}I_t \mathrm{d}Q_t = \cos\theta_t \sin\theta_t (\mathrm{d}R_t^2 - R_t^2 \mathrm{d}\theta_t^2). \qquad (8.39)$$

This leads to the following result.

Proposition 8.11 *The I_t, Q_t components of Ψ_t are independent if and only if $\sigma_{(R)} = R\sigma_{(\theta)}$, i.e. $\sigma_{(z)} = 2z\sigma_{(\theta)}$. A departure from this relation induces a correlation between the Wiener increments in I_t, Q_t.*

Alternatively, this result can be derived using the contravariance of the diffusion tensor σ^{ij}, which enables one to translate between its I, Q and R, θ components via the above coordinate transformation (e.g. Risken 1989). In the general case, we find from (8.20), (8.30) that

$$\frac{\sigma^2_{(z)}}{\sigma^2_{(\theta)}} = 4z^2 + \frac{4\mathcal{A}z^3}{\mathcal{B}x^2}. \tag{8.40}$$

This relation can be used to characterize the geometry of the fluctuations as follows.

Proposition 8.12 *In the K-distributed case, $\mathcal{A} \neq 0$, the diffusion tensor is non-degenerate, and the fluctuations δI_t, δQ_t are correlated. The (comoving) error surface S of $\delta\Psi_t$, defined by the quadratic form*

$$\sigma^{II}\delta I_t^2 + 2\sigma^{IQ}\delta I_t \delta Q_t + \sigma^{QQ}\delta Q_t^2 = 1, \tag{8.41}$$

is an ellipse whose major axis lies in the instantaneous radial direction defined by Ψ_t. Degeneracy occurs only in the Rayleigh case, $\mathcal{A} = 0$, for which S is a circle, i.e. the fluctuations in Ψ_t are isotropic.

We remark, in general, that the random variables I_t, Q_t possess a joint probability distribution that is $U(1)$ symmetric, i.e. given by a surface of revolution about the perpendicular axis to the origin in the I, Q-plane. Nevertheless I_t, Q_t are correlated in general, and become independent only in the Rayleigh case, $\mathcal{A} = 0$, for which the surface of revolution is Gaussian. In this case the component I_t, Q_t processes can be described by the pair of (uncoupled) Ornstein–Uhlenbeck processes determined as the real and imaginary parts of (6.21).

8.3 Asymptotic behaviour

In this section we shall investigate the equilibrium and detailed balance properties of K-scattering. The mathematical tools involved are the FPE and the concept of stochastic current as introduced in Section 3.2.

8.3.1 Equilibrium distribution

We recall the covariant form of the FPE for the asymptotic joint distribution $\mathcal{P}(x, z, t)$ (e.g. Risken 1989)

$$\frac{\partial \mathcal{P}}{\partial t} = -\sum_i \partial_i(b^i \mathcal{P}) + \frac{1}{2}\sum_{i,j} \partial_i \partial_j (\sigma^{ij}\mathcal{P}). \tag{8.42}$$

From (8.9), (8.18), (8.19) the components of the diffusion tensor in the x, z coordinate representation are given by

$$\sigma^{ij} = 2 \begin{pmatrix} \mathcal{A}x & \mathcal{A}z \\ \mathcal{A}z & \mathcal{B}xz + \mathcal{A}z^2/x \end{pmatrix} \tag{8.43}$$

while the drift vector has components

$$\beta^i = 2 \begin{pmatrix} \mathcal{A}(\alpha - x) \\ \mathcal{B}(x - z) + \mathcal{A}z(\alpha - x)/x \end{pmatrix}. \tag{8.44}$$

From (8.42), therefore, we deduce the following.

Proposition 8.13 *The FPE for the joint distribution of the cross-section x_t and intensity z_t is*

$$\begin{aligned}
\frac{\partial \mathcal{P}}{\partial t} &= \mathcal{B}\left\{-\partial_z\left[(x-z)\mathcal{P}\right] + x\partial_z^2\left[z\mathcal{P}\right]\right\} \\
&+ \mathcal{A}\left\{-\partial_z\left(\frac{z(\alpha-x)\mathcal{P}}{x}\right) - \partial_x((\alpha-x)\mathcal{P})\right\} \\
&+ \mathcal{A}\left\{\partial_z^2\left(\frac{z^2\mathcal{P}}{x}\right) + 2\partial_x\partial_z(z\mathcal{P}) + \partial_x^2(x\mathcal{P})\right\}.
\end{aligned} \tag{8.45}$$

This admits the asymptotic joint distribution

$$\mathcal{P} = \frac{x^{\alpha-2}\exp(-x - z/x)}{\Gamma(\alpha)}. \tag{8.46}$$

Proof The derivation of (8.45) follows immediately from (8.42), (8.43), and (8.44), while the following identities for the derivatives of the joint distribution

$$\begin{aligned}
\partial_z \mathcal{P} &= -\frac{\mathcal{P}}{x} \\
\partial_z(z\mathcal{P}) &= \left(1 - \frac{z}{x}\right)\mathcal{P} \\
\partial_z^2(z\mathcal{P}) &= \left(-\frac{2}{x} + \frac{z}{x^2}\right)\mathcal{P} \\
\partial_x \mathcal{P} &= \left(\frac{\alpha-1}{x} + \frac{z}{x^2} - 1\right)\mathcal{P} \\
\partial_x^2 \mathcal{P} &= \left\{\left(\frac{\alpha-1}{x} + \frac{z}{x^2} - 1\right)^2 - \frac{\alpha-1}{x^2} - \frac{2z}{x^3}\right\}\mathcal{P}
\end{aligned} \tag{8.47}$$

enable one to verify that (8.46) is an asymptotic solution of (8.45).

Observe from (8.9) and (8.45) that the model behaviour of the scattering cross-section is *endogenously specified*, i.e. the parameters involved arise from the population alone, independently of the electromagnetic field. Thus (8.9) is independent of z_t. Nevertheless, there exists a non-linear coupling between the x_t, z_t variables, owing to the presence of x_t in (8.18) and the correlation of $W_t^{(x)}$, $W_t^{(z)}$ according to (8.19). Therefore x_t, z_t are statistically dependent random variables, which relation is symmetric. The situation in regard to the endogenous specification of the evolution of the cross-section through (8.9) should be contrasted with the previous discussion given in Tough (1987), and its generalizations in Field and Tough (2003a) in which it is necessary that the SDE for x_t has an explicit z_t dependence, and in which the Wiener processes $W_t^{(x)}$, $W_t^{(z)}$ are considered to be independent. Although these analyses preserve the joint distribution appropriate to K-distributed noise, these models are not so natural from a physical point of view.

8.3.2 Detailed balance

The covariant FPE (8.42) can be re-expressed as the equation of continuity

$$\frac{\partial \mathcal{P}}{\partial t} + \sum_i \partial_i(\mathcal{P} v^i) = 0 \qquad (8.48)$$

where the *current* v^i is defined by

$$v^i = b^i - \frac{1}{2}\mathcal{P}^{-1} \sum_j \partial_j(\sigma^{ij}\mathcal{P}). \qquad (8.49)$$

In addition to the equilibrium condition $\partial \mathcal{P}/\partial t = 0$, the condition for detailed balance states that $v^i = 0$. Explicit calculation using (8.47) shows that v^i vanishes asymptotically. Alternatively, a more intuitive argument for this property is as follows. From (8.3), we have the factorization $z_t = x_t u_t$, where $u_t = |\gamma_t|^2$, in which the factors x_t, u_t are independent random variables. The coordinate transformation $x^{\mathbf{i}} \mapsto x^{\mathbf{i}'} : (x,z) \mapsto (x,u)$ recasts the joint distribution (8.46) and SDEs (8.9), (8.18) such that

$$\beta^{\mathbf{i}'} = \begin{pmatrix} \mathcal{A}\beta^{\mathbf{1}'}(x) \\ \mathcal{B}\beta^{\mathbf{2}'}(u) \end{pmatrix}, \qquad (8.50)$$

$$\sigma^{\mathbf{i}'\mathbf{j}'} = \begin{pmatrix} \mathcal{A}\Sigma^{\mathbf{1}'\mathbf{1}'}(x) & 0 \\ 0 & \mathcal{B}\Sigma^{\mathbf{2}'\mathbf{2}'}(u) \end{pmatrix} \qquad (8.51)$$

where the functions $\beta^{\cdot}(\cdot)$, $\Sigma^{\cdot}(\cdot)$ are determined from (8.4), (8.9) and are independent of $\mathcal{A}$, $\mathcal{B}$. The equilibrium condition in the (x,u) representation, obtained by setting the left-hand side of (8.42) equal to zero, implies detailed balance, since this condition holds for arbitrary values of the constants $\mathcal{A}$, $\mathcal{B}$. Consequently, v^i in the (x,z) representation also vanishes, since v^i transforms homogeneously (i.e. tensorially) under coordinate transformations (see e.g. Risken 1989).

8.4 Correlation and spectra

For simplicity, we adopt a timescale such that the constant $\mathcal{B}$ of (8.4) is equal to unity. The independent constant $\mathcal{A}$ will then satisfy $\mathcal{A} \ll 1$ in most practical situations of interest (e.g. scattering at radar wavelength), although this condition is not necessary for the validity of the expressions that follow in this section.

8.4.1 Intensity autocorrelation

It is convenient to write the intensity process in the product representation $z_t = u_t x_t$. From (8.4) the process $u_t = |\gamma_t|^2$ satisfies the SDE

$$ du_t = (1 - u_t)dt + \sqrt{2u_t}dW_t^{(u)}, \qquad (8.52) $$

where $\gamma_t d\xi_t^* + \gamma_t^* d\xi_t = \sqrt{2u_t}dW_t^{(u)}$. The propagator (i.e. Green's function for the corresponding FPE) for the process u_t is given by

$$ P(u,t|u_0) = \frac{1}{1-\exp(-t)} \exp\left(-\frac{u + u_0 \exp(-t)}{1 - \exp(-t)}\right) I_0\left(\frac{2\exp(-t/2)\sqrt{uu_0}}{1 - \exp(-t)}\right) \qquad (8.53) $$

where I_α denotes the modified Bessel function (e.g. Jeffreys and Jeffreys 1966). In a similar manner, the propagator for (8.9) is given by

$$ P(x,t|x_0) = \frac{1}{1-\exp(-\mathcal{A}t)} \left(\frac{x\exp(\mathcal{A}t)}{x_0}\right)^{(\alpha-1)/2} \exp\left(-\frac{(x + x_0\exp(-\mathcal{A}t))}{1-\exp(-\mathcal{A}t)}\right) $$
$$ \times I_{\alpha-1}\left(\frac{2\exp(-\mathcal{A}t/2)\sqrt{xx_0}}{1-\exp(-\mathcal{A}t)}\right). \qquad (8.54) $$

This can be re-expressed as a series expansion

$$ P(x,t|x_0) = x^{\alpha-1}\exp(-x)\sum_{n=0}^{\infty} \frac{n!}{\Gamma(n+\alpha)} \exp(-\mathcal{A}nt) L_n^{\alpha-1}(x) L_n^{\alpha-1}(x_0), \qquad (8.55) $$

where the Laguerre polynomials L_n^α are defined by

$$ L_n^\alpha(x) = \frac{x^{-\alpha}\exp(x)}{n!} \left(\frac{d}{dx}\right)^n (x^{\alpha+n}\exp(-x)) \qquad (8.56) $$

(cf. Wong 1963 for corresponding derivations). Combining (8.53), (8.55) leads to the following result.

Proposition 8.14 *The propagator for (8.45) is given by*

$$P(z,x,t|z_0,x_0)$$
$$= \frac{1}{x(1-\exp(-t))(1-\exp(-\mathcal{A}t))}\left(\frac{x\exp(\mathcal{A}t)}{x_0}\right)^{(\alpha-1)/2}$$
$$\times \exp\left(-\frac{z/x+z_0\exp(-t)/x_0}{1-\exp(-t)}\right)\exp\left(-\frac{x+x_0\exp(-\mathcal{A}t)}{1-\exp(-\mathcal{A}t)}\right)$$
$$\times I_0\left(\frac{2\exp(-t/2)}{1-\exp(-t)}\sqrt{\frac{zz_0}{xx_0}}\right)I_{\alpha-1}\left(\frac{2\exp(-\mathcal{A}t/2)\sqrt{xx_0}}{1-\exp(-\mathcal{A}t)}\right). \quad (8.57)$$

Thus a general two-point correlation function can be expressed as the integral

$$\langle F_1(x_t,z_t)F_2(x_0,z_0)\rangle = \int_0^\infty dx\,dz\,dx_0\,dz_0\,F_1(x,z)F_2(x_0,z_0)P(x,z,t|x_0,z_0)$$
$$\times \frac{x_0^{\alpha-2}\exp(-z_0/x_0-x_0)}{\Gamma(\alpha)}. \quad (8.58)$$

In particular, we deduce the following important consequence.

Corollary 8.15 *The intensity autocorrelation function is determined by the following expression:*

$$\langle z_t z_0\rangle = \langle u_t u_0\rangle\langle x_t x_0\rangle$$
$$= \alpha(\alpha + \exp(-\mathcal{A}t))(1+\exp(-t)). \quad (8.59)$$

8.4.2 Power spectral density

In the additional presence of a Doppler frequency shift ω_0[22], the process γ_t of Section 6.2 is modified to obey the SDE

$$d\gamma_t = \left(-\frac{1}{2}+i\omega_0\right)\gamma_t dt + d\xi_t. \quad (8.60)$$

The amplitude process Ψ_t determined by (8.12) is *stationary*, since there is no explicit time dependence in (8.12), the phase distribution is uniform, and the modulus amplitude R_t has a stationary distribution in accordance with the stationary K-distribution for the intensity (it is assumed that the distributions of the *initial* values of Ψ and its associated processes are given by their asymptotic stationary distributions). Therefore, we apply the *Wiener–Khintchine theorem* which asserts that the *power spectral density* $S(\omega)$ is equal to the Fourier transform (denoted tilde) of the autocorrelation function, i.e.

$$\langle\tilde\Psi(\omega)\tilde\Psi(\omega')\rangle = \pi\delta(\omega-\omega')S(\omega), \quad (8.61)$$

where $S(\omega) = \langle\widetilde{\Psi_t\Psi_0^*}\rangle$.

---

[22]The presence of $\omega_0 \neq 0$ is important in radar applications; see e.g. Helstrom 1960.

Remarks. *Concerning power spectra and stationarity.* In the case of a nonstationary process the power spectral density is defined as in the left-hand side of (8.61), i.e. the autocorrelation function in the frequency domain – thus, in general, it is a function of a *pair* of frequencies. The stationarity property leads to the delta function dependence on the right-hand side of (8.61) so that, in this case, the power spectral density becomes essentially a function of a single frequency only.

The amplitude autocorrelation function satisfies

$$\langle \Psi_t \Psi_0^* \rangle = \langle \sqrt{x_t x_0} \rangle \exp(-|t|/2 - i\omega_0 t). \tag{8.62}$$

Using the propagator expansion (8.55), the evaluation of the factor $\langle \sqrt{x_t x_0} \rangle$ proceeds according to

$$\langle \sqrt{x_t x_0} \rangle = \int_0^\infty dx \int_0^\infty dx_0 \frac{x_0^{\alpha-1} \exp(-x_0)}{\Gamma(\alpha)} P(x,t|x_0) \sqrt{x x_0}$$

$$= \frac{1}{\Gamma(\alpha)} \sum_{n=0}^\infty \frac{n!}{\Gamma(n+\alpha)} \exp(-\mathcal{A}nt) \left(\int_0^\infty x^{\alpha-\frac{1}{2}} \exp(-x) L_n^{\alpha-1}(x) dx \right)^2$$

$$= \frac{1}{\Gamma(\alpha)} \sum_{n=0}^\infty \frac{n!}{\Gamma(n+\alpha)} \exp(-\mathcal{A}nt) \left(\frac{\Gamma(\alpha+\frac{1}{2})\Gamma(n-\frac{1}{2})}{n! 2\sqrt{\pi}} \right)^2$$

$$= \frac{\Gamma(\alpha+\frac{1}{2})^2}{\Gamma(\alpha)^2} {}_2F_1\left(-\frac{1}{2}, -\frac{1}{2}, \alpha, \exp(-\mathcal{A}t)\right). \tag{8.63}$$

Here the hypergeometric function ${}_2F_1$ is identified from its series expansion. When $t \to \infty$, (8.63) approaches $\langle \sqrt{x} \rangle^2$, as anticipated from the decorrelation of x_t, x_0 over large times. As $t \to 0$ we find, from the identity due to Gauss,

$${}_2F_1(a,b,c,1) = \frac{\Gamma(c)\Gamma(c-a-b)}{\Gamma(c-a)\Gamma(c-b)}, \tag{8.64}$$

that (8.63) reduces to the anticipated form

$$\lim_{t \to 0} \langle \sqrt{x_t x_0} \rangle = \frac{\Gamma(\alpha+1)}{\Gamma(\alpha)} = \alpha = \langle x \rangle. \tag{8.65}$$

Expressions (8.62), (8.63) lead to the following result.

Proposition 8.16 *The autocorrelation function of the complex amplitude process Ψ_t is given by*

$$\langle \Psi_t \Psi_0^* \rangle = \frac{\Gamma(\alpha+\frac{1}{2})^2}{\Gamma(\alpha)^2} {}_2F_1\left(-\frac{1}{2}, -\frac{1}{2}, \alpha, \exp(-\mathcal{A}t)\right) \exp\left(-\frac{|t|}{2} - i\omega_0 t\right). \tag{8.66}$$

According to the Wiener–Khintchine theorem, a Fourier transform of this result has the following consequence.

Corollary 8.17 *The power spectral density of the K-distributed noise process characterized by (8.12) is given by*

$$S(\omega) = 2\frac{\Gamma(\alpha+\frac{1}{2})^2}{\Gamma(\alpha)^2}\int_0^\infty {}_2F_1\left(-\frac{1}{2},-\frac{1}{2},\alpha,\exp(-\mathcal{A}t)\right)\exp\left(-\frac{t}{2}\right)\cos((\omega-\omega_0)t). \tag{8.67}$$

Expanding the hypergeometric function ${}_2F_1$ *as a series and integrating term by term, the resulting series is recognized as a generalized hypergeometric function of unit argument. Thus*

$$S(\omega)$$
$$= 2\Re\left\{\frac{{}_3F_2(-1/2,-1/2,(1/2+i(\omega-\omega_0))/\mathcal{A};\alpha,1+(1/2+i(\omega-\omega_0))/\mathcal{A};1)}{1/2+i(\omega-\omega_0)}\right\}$$
$$\times \frac{\Gamma(\alpha+\frac{1}{2})^2}{\Gamma(\alpha)^2}. \tag{8.68}$$

These calculations illustrate how the compound representation of the amplitude (8.3) facilitates the analysis of the associated FPE (8.45). In terms of the constituent spectra for the two component factors in (8.3) observe that, since these component processes are independent, the autocorrelation of the resultant amplitude factorizes into that of the components. Therefore, according to the Wiener–Khintchine and convolution theorems, the power spectrum of the resultant amplitude is equal to the convolution of the component spectra.

The explicit dependence on the electromagnetic frequency scale $\mathcal{B}$ in the above expressions may be restored most simply on dimensional grounds.[23] The frequency scales $\mathcal{A}$ and $\mathcal{B}$, which satisfy $\mathcal{A} \ll \mathcal{B}$ (for typical carrier frequencies), may be found from an experimentally observed autocorrelation by fitting these parameters according to the theoretical expression (8.66). From the series expansion of the hypergeometric function, the radar cross-section (RCS) component of the resultant amplitude autocorrelation (8.66) may be written as a sum of terms proportional to $\exp(-n\mathcal{A}t)$ (Fayard 2008), whose spectra are therefore Cauchy or 'Lorentzian', while the Rayleigh spectral component is also Cauchy.[24] Since the Cauchy distribution is stable (see Appendix A), via the Fourier convolution $S^{(\psi)} = S^{(r)} * S^{(\gamma)}$ and taking the leading ($n=1$) term in the hypergeometric expansion, it follows that the spectrum of the resultant amplitude ψ is also (approximately) Cauchy, with FWHM equal to $2\mathcal{A}+\mathcal{B}$. (The DC part of the RCS spectrum, equal to $\alpha\delta(\omega)$ and reflecting merely the fact that the RCS has a constant non-zero mean value of α, has been removed for clarity.)

Observe with regard to the above analysis the special case that the cross-section is constant ($\mathcal{A}=0$). Then the amplitude autocorrelation consists solely

---

[23] Thus, for example, a term $\exp(-\frac{1}{2}|t|)$ becomes $\exp(-\frac{1}{2}\mathcal{B}|t|)$ in general units of time, etc..

[24] An autocorrelation function $R(\tau) = \exp(-\frac{1}{2}kt)$ has associated power spectrum $S(\omega) = 2k/\pi(k^2+4\omega^2)$ with full-width-half-maximum (FWHM) therefore equal to k.

of an exponential decay term and thus the spectrum is exactly *Lorentzian*, i.e. a Cauchy distribution (see Appendix A).

8.5 Interpretation and implications

The study provides the first theoretical account of K-distributed noise processes in which the continuous time dynamical features of the electromagnetic scattering process are fully captured. This has been achieved via the formulation of SDEs for the scattered amplitude, using the primitive assumptions of the complex random walk model.

The results substantiate an earlier proposal for anomaly detection in the context of such processes (Field and Tough 2003a) based on the concept of observability in the fluctuations of the complex amplitude process over a sample path. In this respect, Corollary 8.2 leads to a correlation between the observed $|d\Psi_t|^2$ and its predicted value of the form $c(|d\Psi_t|^2, z_t)$, which should approximate unity within the domain of validity of the model. This feature enables an anomaly detection mechanism for K-distributed noise processes, which has been successfully tested on experimental data, and is described later in Section 12.2 herein (cf. §4(*b*) of Field and Tough (2003a) where the result first appears). It is of considerable importance that (8.13) can be derived from theoretical considerations alone, as described in Section 8.1.

The formulation of the continuous time dynamics is more fundamental than knowledge of certain statistical properties of a model and, moreover, implies the form of all correlation functions and higher order statistics. In this respect we have provided closed form expressions for the intensity autocorrelation function and the power spectral density, which should be applicable in situations of radar and laser physics. The tractability of such expressions is facilitated by the use of computational tools such as *Mathematica* (Wolfram 1999).

The methodology we describe admits the generalization of the SDE for the scattering cross-section (8.9) to more general endogenous models of population processes, such as described in *orig.* §2 of Field and Tough (2003a) and Section 7.3 herein. This could include corresponding descriptions of the electromagnetic scattering processes that lead e.g. to the Weibull distribution (applied to scatter from land clutter), the intensity compound K-distribution for various radar parameters (e.g. applied to synthetic aperture radar), and other examples (cf. Jakeman and Tough 1988).

Corollary 8.7 implies that the instantaneous value of the scattering cross-section is observable from the scattered amplitude. The cross-section is of primary significance in anomaly detection within a random scattering medium, e.g. the sea-surface, heterogeneous media, which had previously been regarded as a hidden physical variable whose instantaneous values were not deducible from the scattering data. Expression (8.22) for the squared volatility in the modulus amplitude should also find application to problems in incoherent radar detection where the total scattered phase information is not available.

We have recalled in this chapter that the random walk model with step number fluctuations, due to Jakeman (see Jakeman 1980; Jakeman and Tough 1988), accounts for certain statistical properties of K-scattering. In addition, we have provided the extension to a complete dynamical description, in terms of continuous time diffusion processes. This dynamical extension is generalized further in Chapter 9, where we demonstrate how to include the effect of weak scattering in superposition with a coherent offset signal, in a corresponding stochastic dynamical framework.

9

MODELS OF WEAK SCATTERING

In this chapter, we extend the K-scattering model of Chapter 8, in which the continuous time dynamics of the K-scattering process were derived, to include the effect of the presence of a coherent offset or 'signal' in the scattering amplitude. The weak scattering amplitudes are characterized in terms of continuous time biased random walk models, and the corresponding stochastic dynamics derived. The stochastic differential geometry of the resultant amplitude fluctuations is derived in relation to that of pure K-scattering. Asymptotic distributions of amplitude, intensity and phase are provided, and the condition for detailed balance shown to hold.

Significant progress has recently been made in our understanding of the dynamics of models of electromagnetic scattering in the context of diffusion processes. Deviations from Rayleigh (Gaussian) scattering have been successfully formulated in the context of K-distributed scattering processes (Field and Tough 2003b) and have formed the basis of an anomaly detection technique that has been successfully applied to maritime radar scattering and laser propagation experiments (Field and Tough 2003a).

The models considered previously assume a uniform (asymptotic) distribution of phase. In this chapter, we consider how anisotropic phase distributions can be accommodated within the framework provided by stochastic differential equations (SDEs) that has proved to be useful in K-scattering. We have seen how a simple random walk model provides a physically motivated description of the scattering process (Jakeman and Tough 1988) that at the same time makes useful contact with the SDE formulation of the problem (Field and Tough 2003a,b). In earlier work, un-biased random walk models have provided useful insight into the Gaussian and non-Gaussian statistics of radiation scattered sufficiently strongly for its phase to be effectively randomized, and to take a uniform asymptotic distribution. A biased random walk model of weak scattering has been discussed in detail in Jakeman and Tough (1987). Their analysis led to the so-called generalized K-scattering model. The present chapter re-addresses this problem, replacing the static, characteristic function, approach with one in which the dynamics is captured by a set of coupled SDEs. A fairly complete analysis is possible that also makes contact with the Rice and homodyned K descriptions of weak scattering (Jakeman 1980). This allows for a detailed description of the geometry of the resultant amplitude fluctuations, which is shown to be different in some significant respects from that encountered in the K-distributed case (Field and Tough 2003b). In addition to developing this SDE description, we study the phase distributions implicit in these models in more detail than has been reported previously.

It will be necessary for the reader to review the results of Chapter 8 (*orig.* Field and Tough 2003*b*), which are essential in the present context for the treatment of weak scattering.

9.1 Weak scattering amplitudes

In situations of strong back-scattering, such as that occur e.g. in radar applications, the phases of the back-scattered components are taken to be uniformly randomized and correspondingly the dynamics and asymptotic distributions of the resultant amplitude process Ψ_t are invariant under $\Psi_t \mapsto e^{i\Lambda}\Psi_t$. This is no longer the case for 'weak' scattering however, i.e. situations where the Rayleigh component of the scatter is weak in comparison to some coherent offset contribution. In these cases the mean amplitude is offset from zero, and the asymptotic resultant phase distribution is anisotropic. We have seen in the K-distributed case that $\psi_t = x_t^{1/2} \gamma_t$ where γ describes the (unit-power) Rayleigh process (Field and Tough 2003*b*) according to (8.4). When this process lies in superposition with a coherent offset amplitude ϱ_t, the resultant amplitude process Ψ_t depends on the relative scalings of the offset and (modulated) Rayleigh components with respect to population size. There are essentially three cases to consider, each of which can be understood in terms of the random walk model (6.16) by imposing a *bias* on each step $s^{(j)}$, whose physical origin is the coherent offset contribution. We shall describe these cases in the order of Rice, homodyned, and generalized K-scattering, thus introducing physical features (noise, K-noise, fluctuating coherently scattered beam) in a natural order that is mirrored in the increasing complexity of the calculations.

9.1.0.1 *Rice* We assume that the number of scatterers is constant in time, with a constant offset contribution $\varrho_t = a$. Thus modifying the random walk model (6.16) we write

$$\mathcal{E}_t^{(N)} = \sum_{j=1}^{N} \left(\overbrace{a + \exp[i\varphi_t^{(j)}]}^{s^{(j)}} \right). \tag{9.1}$$

Scaling by $1/N$, $1/N^{\frac{1}{2}}$ for the respective terms under the summation, in the x_t-continuum limit ($N \to \infty$) this becomes

$$\Psi_t^{\text{R}} = a + \gamma_t. \tag{9.2}$$

9.1.0.2 *Homodyned K* The situation here is the same as for K-scattering with the superposition of a constant offset $\varrho_t = a$ that does not fluctuate with N_t. In the continuum limit this amounts to adding a constant to the K-amplitude, thus

$$\Psi_t^{\text{HK}} = a + \psi_t. \tag{9.3}$$

9.1.0.3 *Generalized K* In a similar fashion (9.1) is modified to become

$$\mathcal{E}_t^{(N_t)} = \sum_{j=1}^{N_t} \left(\overbrace{a + \exp[i\varphi_t^{(j)}]}^{s^{(j)}} \right). \tag{9.4}$$

in which the t-dependence of the limit of summation is to be observed, i.e. the step number of the biased random walk has fluctuations. Scaling by the reciprocal mean and root mean populations respectively, the offset becomes $\varrho_t = ax_t$ and we have

$$\Psi_t^{\text{GK}} = ax_t + \psi_t \tag{9.5}$$

in the continuum limit. Observe with respect to scaling in the continuum population limit that, in each case, we have divided by the (unique) length scale factors, appropriate to the relevant terms in $s^{(j)}$ separately, which yield finite non-zero resultant amplitudes.

9.2 Stochastic dynamics

The stochastic dynamics of the weak scattering amplitudes described above can be calculated from the underlying K-scattering dynamics presented in Chapter 8. We shall make use of the identities for the (resultant) intensity and phase stochastic differentials in terms of the (resultant) amplitude,

$$dZ_t \equiv \Psi_t^* d\Psi_t + \Psi_t d\Psi_t^* + |d\Psi_t|^2,$$

$$d\Theta_t \equiv \frac{1}{2i} \left[\left(\frac{d\Psi_t}{\Psi_t} - \frac{1}{2} \left(\frac{d\Psi_t}{\Psi_t} \right)^2 \right) - \left(\frac{d\Psi_t^*}{\Psi_t^*} - \frac{1}{2} \left(\frac{d\Psi_t^*}{\Psi_t^*} \right)^2 \right) \right], \tag{9.6}$$

and their products

$$\Sigma_t^{(Z)} \equiv \Psi_t^2 \Sigma_t^{(\Psi^*)} + \Psi_t^{*2} \Sigma_t^{(\Psi)} + 2Z_t \Sigma_t^{(\Psi,\Psi^*)},$$

$$\Sigma_t^{(Z,\Theta)} \equiv \Im \left[\left(\frac{\Psi_t^*}{\Psi_t} \right) \Sigma_t^{(\Psi)} \right], \tag{9.7}$$

$$\Sigma_t^{(\Theta)} \equiv \frac{1}{4} \left[\frac{2\Sigma_t^{(\Psi,\Psi^*)}}{Z_t} - \frac{\Sigma_t^{(\Psi)}}{\Psi_t^2} - \frac{\Sigma_t^{(\Psi^*)}}{\Psi_t^{*2}} \right].$$

In combination with (9.2), (9.3), (9.5), (8.9) and the results of Corollary 8.9 these identities enable us to derive the SDEs satisfied by Z_t, Θ_t in terms of the component Wiener processes $\{W_t^{(r)}, W_t^{(\theta)}, W_t^{(x)}\}$ encountered in Chapter 8. The dynamics are simplest for Rice scattering owing to the differential of (9.2). (In the context of radar applications, the Rice scattering model is referred to as a 'Swerling zero target in Rayleigh clutter', where the 'target' strength is represented by the signal ϱ (assumed constant over the timescale of interest) and the Rayleigh process γ_t represents background 'clutter'.) More care is required in the calculations for the homodyned and generalized K-scattering processes owing to certain cross-terms that arise. Nevertheless the strategy is the same for each case, and we are led to the dynamical characterizations of the vector scattering process $\mathbf{S}_t = (x_t, Z_t, \Theta_t)^{\text{tr}}$ according to the scheme

$$dS_t^i = \beta_t^i dt + \sigma_t^i dW_t^i \tag{9.8}$$

(no summation over i) for a collection of Wiener processes $\{W_t^i \mid \forall i\}$ (not necessarily independent) with respective drift and diffusion coefficients β^i, Σ^{ij} determined by

$$\beta_t^i = \frac{\mathbf{E}[dS_t^i]}{dt}, \tag{9.9}$$
$$dS_t^i dS_t^j = \Sigma_t^{ij} dt.$$

The corresponding Fokker–Planck equation (FPE) (e.g. Risken 1989) for the joint probability density $\rho_t(x, Z, \Theta)$ is then

$$\partial_t \rho + \partial_i (\rho \mathcal{V}^i) = 0 \tag{9.10}$$

where the vector scattering current $\mathcal{V}^i$ is defined by

$$\mathcal{V}^i = \beta^i - \frac{1}{2}\rho^{-1}\partial_j(\Sigma^{ij}\rho). \tag{9.11}$$

9.2.0.4 Rice
The amplitude dynamics of the Rice process is identical to that of the Rayleigh process and the cross-section is constant and equal to unity, as evident from (9.2). We deduce from the identities above that, in terms of the geometry of the underlying Rayleigh process, the resultant intensity satisfies the SDE

$$dZ_t = \mathcal{B}\left[1 - u_t - au_t^{\frac{1}{2}}\cos\theta_t\right]dt + (2\mathcal{B})^{\frac{1}{2}}\left[(u_t^{\frac{1}{2}} + a\cos\theta_t)dW_t^{(r)} - a\sin\theta_t dW_t^{(\theta)}\right]. \tag{9.12}$$

Likewise the resultant phase satisfies

$$d\Theta_t = \frac{-\frac{1}{2}\mathcal{B}au_t^{1/2}\sin\theta_t dt + \left(\frac{\mathcal{B}}{2}\right)^{1/2}\left[a\sin\theta_t dW_t^{(r)} + (u_t^{1/2} + a\cos\theta_t)dW_t^{(\theta)}\right]}{\left(a^2 + u_t + 2au_t^{1/2}\cos\theta_t\right)}. \tag{9.13}$$

This leads to the following result.

Proposition 9.1 *The Rice vector scattering process* $\mathbf{S}_t$ *has drift*

$$\beta^i = \begin{pmatrix} \mathcal{B}\left[1 - Z + aZ^{\frac{1}{2}}\cos\Theta\right] \\ -(\mathcal{B}a\sin\Theta)/2Z^{\frac{1}{2}} \end{pmatrix}, \qquad (9.14)$$

and diffusion tensor

$$\Sigma^{ij} = \begin{pmatrix} 2\mathcal{B}Z & 0 \\ 0 & \mathcal{B}/2Z \end{pmatrix}. \qquad (9.15)$$

9.2.0.5 Homodyned K From (9.3) the amplitude dynamics is identical to that of the K-process. Thus for the intensity, in terms of the underlying K-scattering geometry, we find

$$dZ_t = \left[\frac{\mathcal{A}z_t(\alpha - x_t)}{x_t} + \mathcal{B}(x_t - z_t) + az_t^{\frac{1}{2}}\cos\theta_t\left(\mathcal{A}\left[\frac{\alpha - x_t - \frac{1}{2}}{x_t}\right] - \mathcal{B}\right)\right]dt$$

$$+ (2\mathcal{B}x_t)^{\frac{1}{2}}\left[a\cos\theta_t + z_t^{\frac{1}{2}}\right]dW_t^{(r)} - (2\mathcal{B}x_t)^{\frac{1}{2}}[a\sin\theta_t]dW_t^{(\theta)}$$

$$+ \left(\frac{2\mathcal{A}}{x_t}\right)^{\frac{1}{2}}\left[az_t^{\frac{1}{2}}\cos\theta_t + z_t\right]dW_t^{(x)}. \qquad (9.16)$$

Likewise for the resultant phase we find

$$d\Theta_t = \frac{az_t^{\frac{1}{2}}\sin\theta_t}{Z_t}\left\{\mathcal{A}\left[\left(\frac{2(\alpha - x_t) - 1}{4x_t}\right) - \frac{z_t^{\frac{1}{2}}}{2x_t Z_t}\left(z_t^{\frac{1}{2}} + a\cos\theta_t\right)\right] - \frac{1}{2}\mathcal{B}\right\}dt$$

$$+ \left(\frac{\mathcal{B}x_t}{2}\right)^{\frac{1}{2}}\frac{a\sin\theta_t}{Z_t}dW_t^{(r)} + \left(\frac{\mathcal{B}x_t}{2}\right)^{\frac{1}{2}}\frac{z_t^{\frac{1}{2}} + a\cos\theta_t}{Z_t}dW_t^{(\theta)}$$

$$+ \left(\frac{\mathcal{A}z_t}{2x_t}\right)^{\frac{1}{2}}\frac{a\sin\theta_t}{Z_t}dW_t^{(x)}. \qquad (9.17)$$

Thus, in terms of the resultants (Z, Θ), we deduce that the vector homodyned K-scattering process (x_t, Z_t, Θ_t) has the following structure.

Proposition 9.2 *The drift vector is given by*

$$\beta^i = \begin{pmatrix} \mathcal{A}(\alpha - x) \\ \frac{\mathcal{A}}{x}\left[(\alpha - x)Z + a\left(x - \alpha - \frac{1}{2}\right)Z^{\frac{1}{2}}\cos\Theta + \frac{1}{2}a^2\right] + \mathcal{B}(x - Z + aZ^{\frac{1}{2}}\cos\Theta) \\ \frac{a\sin\Theta}{Z^{\frac{1}{2}}}\left\{\mathcal{A}\left[\left(\frac{2(\alpha - x) - 1}{4x}\right) - \frac{1}{2xZ^{\frac{1}{2}}}(Z^{\frac{1}{2}} - a\cos\Theta)\right] - \frac{1}{2}\mathcal{B}\right\} \end{pmatrix}. \qquad (9.18)$$

The (symmetric) diffusion tensor is

$$\Sigma^{ij} = \begin{pmatrix} 2\mathcal{A}x & 2\mathcal{A}(Z - aZ^{1/2}\cos\Theta) & \frac{\mathcal{A}a\sin\Theta}{Z^{1/2}} \\ \ldots & 2Z\left[\frac{\mathcal{A}(Z^{1/2} - a\cos\Theta)^2}{x} + \mathcal{B}x\right] & (\mathcal{A}a\sin\Theta)(Z^{1/2} - a\cos\Theta)/x \\ \ldots & \ldots & \frac{1}{2Z}\left[\frac{\mathcal{A}a^2\sin^2\Theta}{x} + \mathcal{B}x\right] \end{pmatrix}. \qquad (9.19)$$

9.2.0.6 Generalized K
The differential of the amplitude (9.5) contains both a K-scattering component and an explicit fluctuating part from the cross-section, i.e. $d\Psi_t = a dx_t + d\psi_t$. This leads, in terms of the K-scattering geometry, to the intensity SDE

$$\begin{aligned}
dZ_t =& \mathcal{A}\left[\frac{z_t(\alpha - x_t)}{x_t} + 2a^2 x_t(\alpha - x_t + 1) + 3az_t^{\frac{1}{2}}\left(\alpha - x_t + \frac{1}{2}\right)\cos\theta_t\right]dt \\
&+ \mathcal{B}\left(x_t - z_t - ax_t z_t^{\frac{1}{2}}\cos\theta_t\right)dt + (2\mathcal{B}x_t)^{\frac{1}{2}}\left(z_t^{\frac{1}{2}} + ax_t\cos\theta_t\right)dW_t^{(r)} \\
&- (2\mathcal{B}x_t)^{\frac{1}{2}} ax_t \sin\theta_t dW_t^{(\theta)} + (2\mathcal{A}x_t)^{\frac{1}{2}}\left(\frac{z_t}{x_t} + 2a^2 x_t + 3az_t^{\frac{1}{2}}\cos\theta_t\right)dW_t^{(x)}.
\end{aligned}$$
(9.20)

Likewise for the phase we find

$$\begin{aligned}
d\Theta_t =& \frac{az_t^{\frac{1}{2}} \sin\theta_t}{2Z_t}\left[\mathcal{A}\left(\frac{z_t + 2a^2 x_t^2 + 3ax_t z_t^{\frac{1}{2}}\cos\theta_t}{Z_t} + x_t - \alpha - \frac{1}{2}\right) - \mathcal{B}x_t\right]dt \\
&+ \left(\frac{\mathcal{B}x_t}{2}\right)^{\frac{1}{2}}\frac{ax_t \sin\theta_t}{Z_t}dW_t^{(r)} + \left(\frac{\mathcal{B}x_t}{2}\right)^{\frac{1}{2}}\frac{z_t^{\frac{1}{2}} + ax_t \cos\theta_t}{Z_t}dW_t^{(\theta)} \\
&- \left(\frac{\mathcal{A}x_t z_t}{2}\right)^{\frac{1}{2}}\frac{a \sin\theta_t}{Z_t}dW_t^{(x)}.
\end{aligned}$$
(9.21)

Combining these results we obtain the following.

Proposition 9.3 *The vector generalized K-scattering process has drift*

$$\beta^{\mathbf{i}} = \begin{pmatrix} \mathcal{A}(\alpha - x) \\ \mathcal{A}\left[Z(\alpha/x - 1) + \frac{1}{2}a^2 x + aZ^{\frac{1}{2}}(\alpha - x + \frac{3}{2})\cos\Theta\right] + \mathcal{B}(x - Z + aZ^{\frac{1}{2}}x\cos\Theta) \\ a\sin\Theta\left\{\mathcal{A}\left[(Z^{1/2} + ax\cos\Theta)/Z^{1/2} - (\alpha - x + \frac{1}{2})\right] - \mathcal{B}x\right\}/2Z^{1/2} \end{pmatrix}$$
(9.22)

and (symmetric) diffusion tensor

$$\Sigma^{\mathbf{ij}} = \begin{pmatrix} 2\mathcal{A}x & 2\mathcal{A}(Z + axZ^{1/2}\cos\Theta) & -\mathcal{A}ax\sin\Theta/Z^{1/2} \\ \ldots & 2\mathcal{A}(Z + axZ^{1/2}\cos\Theta)^2/x + 2\mathcal{B}xZ & -\mathcal{A}a(Z^{1/2} + ax\cos\Theta)\sin\Theta \\ \ldots & \ldots & (\mathcal{B} + \mathcal{A}a^2 \sin^2\Theta)x/2Z \end{pmatrix}.$$
(9.23)

9.3 Geometry of amplitude fluctuations

In a similar fashion to Section 8.2, we make contact here between the stochastic differential geometry of Chapter 3 and the detailed (amplitude) dynamics of weak scattering processes.

AMPLITUDE FLUCTUATIONS

We begin with some purely geometrical results concerning the correlation structure in the amplitude fluctuations. Combining drift terms as quantities of $o(\mathrm{d}t^{\frac{1}{2}})$, we write the amplitude stochastic differential as

$$\begin{aligned}
\mathrm{d}\Psi_t &= iR_t \exp(i\Theta_t)\mathrm{d}\Theta_t + \exp(i\Theta_t)\mathrm{d}R_t + o(\mathrm{d}t^{\frac{1}{2}}) \\
&= \alpha_t \exp[i(\Theta_t + \phi_t)] + i\beta_t \exp[i(\Theta_t + \phi_t)] + o(\mathrm{d}t^{\frac{1}{2}}),
\end{aligned} \quad (9.24)$$

where α_t, β_t are real-valued Ito differentials and ϕ_t is chosen so that their Ito product $\alpha_t \beta_t$ vanishes, i.e. the Wiener components of α_t, β_t are statistically independent (see e.g. Karatzas and Shreve 1988). Comparing the two decompositions of $\mathrm{d}\Psi_t$ above, it follows that (neglecting terms of $o(\mathrm{d}t^{\frac{1}{2}})$)

$$\begin{aligned}
\alpha_t \cos \phi_t - \beta_t \sin \phi_t &= \mathrm{d}R_t \\
\alpha_t \sin \phi_t + \beta_t \cos \phi_t &= R_t \mathrm{d}\Theta_t.
\end{aligned} \quad (9.25)$$

Therefore

$$\begin{aligned}
(\alpha_t^2 - \beta_t^2) \sin 2\phi_t &= 2R_t \mathrm{d}R_t \mathrm{d}\Theta_t \\
(\alpha_t^2 - \beta_t^2) \cos 2\phi_t &= \mathrm{d}R_t^2 - R_t^2 \mathrm{d}\Theta_t^2
\end{aligned} \quad (9.26)$$

up to $o(\mathrm{d}t)$. From (9.6),

$$\begin{aligned}
\left(\Sigma_t^{(\alpha)} - \Sigma_t^{(\beta)}\right) \sin 2\phi_t &= \tfrac{1}{2i} \left[\left(\tfrac{\Psi_t^*}{\Psi_t}\right) \Sigma_t^{(\Psi)} - \left(\tfrac{\Psi_t}{\Psi_t^*}\right) \Sigma_t^{(\Psi^*)} \right] \\
\left(\Sigma_t^{(\alpha)} - \Sigma_t^{(\beta)}\right) \cos 2\phi_t &= \tfrac{1}{2} \left[\left(\tfrac{\Psi_t^*}{\Psi_t}\right) \Sigma_t^{(\Psi)} + \left(\tfrac{\Psi_t}{\Psi_t^*}\right) \Sigma_t^{(\Psi^*)} \right].
\end{aligned} \quad (9.27)$$

Also, from (9.25) we find $\alpha_t^2 + \beta_t^2 = \mathrm{d}R_t^2 + R_t^2 \mathrm{d}\Theta_t^2$ so that

$$\Sigma_t^{(\alpha)} + \Sigma_t^{(\beta)} = \Sigma_t^{(\Psi,\Psi^*)}. \quad (9.28)$$

We deduce from (9.26) that

$$\Sigma_t^{(\alpha)} - \Sigma_t^{(\beta)} = \pm \sqrt{\Sigma_t^{(\Psi)} \Sigma_t^{(\Psi^*)}} \quad (9.29)$$

which, combining with (9.28), yields the following result.

Lemma 9.4

$$\begin{aligned}
\Sigma_t^{(\alpha)} &= \tfrac{1}{2}\left(\Sigma_t^{(\Psi,\Psi^*)} \pm \sqrt{\Sigma_t^{(\Psi)} \Sigma_t^{(\Psi^*)}}\right) \\
\Sigma_t^{(\beta)} &= \tfrac{1}{2}\left(\Sigma_t^{(\Psi,\Psi^*)} \mp \sqrt{\Sigma_t^{(\Psi)} \Sigma_t^{(\Psi^*)}}\right).
\end{aligned} \quad (9.30)$$

with $\pm$ corresponding to the major/minor axes of the error surface of the resultant amplitude, respectively.

Observe that (9.29), (9.30) have the appropriate symmetry under interchange $\alpha \longleftrightarrow \beta$. The angle ϕ_t represents a rotation in the geometry of the resultant amplitude fluctuations relative to the case of pure K-scattering, for which $\phi = 0$. From (9.27) this angle is determined as follows.

Lemma 9.5 *The phase rotation ϕ_t, that yields an orthogonal dyad (see Fig. 9.1) associated with independent Wiener increments in the resultant amplitude process Ψ_t, satisfies the geometrical identity*

$$\tan 2\phi_t = \frac{4Z_t \Sigma_t^{(Z,\Theta)}}{\Sigma_t^{(Z)} - 4Z_t^2 \Sigma_t^{(\Theta)}}. \tag{9.31}$$

Equivalently, in terms of the resultant complex amplitude process, we have the geometrical identity

$$\tan 2\phi_t = -\frac{\Im\left[\Psi_t^2 d\Psi_t^{*2}\right]}{\Re\left[\Psi_t^2 d\Psi_t^{*2}\right]}, \tag{9.32}$$

where $\Re$, $\Im$ denote the real and imaginary parts, respectively.

Before applying this geometry to the weak scattering processes described earlier, as a preliminary we give a result which provides the relationship between the structure of the diffusion tensor that arises in the cases of homodyned and generalized K-scattering.

Proposition 9.6 *The transformation $a \mapsto -ax_t$ maps the homodyned to the generalized K-scattering diffusion tensor of the vector scattering process (x_t, Z_t, Θ_t).*

Proof Choose an arbitrary instant of time, labelled $t = 0$. Define

$$\begin{aligned}\Psi_t^{(\mathrm{GK})} &= ax_t + \psi_t \\ \Psi_t^{(\mathrm{HK})} &= -ax_0 + \psi_t'\end{aligned} \tag{9.33}$$

for all $t \geq 0$, coincident at $t = 0$. Thus $\psi_0' = 2ax_0 + \psi_0$ and otherwise ψ_t, ψ_t' are considered independent K-scattering processes. The result is equivalent to the corresponding (complex-valued) vector processes $\left(x_t, \Psi_t^{(\cdot)}\right)$ having the same diffusion tensor, at the chosen instant. The amplitude components are best computed using the complex polarization, i.e.

$$\Sigma^{\ddot{\mathrm{ii}}} dt = \begin{pmatrix} d\Psi_t^2 & d\Psi_t d\Psi_t^* \\ \cdots & d\Psi_t^{*2} \end{pmatrix}. \tag{9.34}$$

The results of Corollary 8.9 and the above relation between ψ_t, ψ_t' at $t = 0$ imply that $d\Psi_0^{(\cdot)2}$ are identical. Likewise the expressions for $|d\Psi_0^{(\cdot)}|^2$ coincide, by virtue of the cosine rule applied to $\triangle PRR'$ of Fig. 9.1. The same method shows that $dx_t d\Psi_t^{(\cdot)}$ are identical at the chosen instant. $\square$

The image point R' has the physical interpretation of a fluctuating canceling beam, π out of phase with the original $\varrho_t(R)$. A result corresponding to Proposition 9.6 does not hold for the vector scattering drift, as evident from comparing Propositions 9.2 and 9.3.

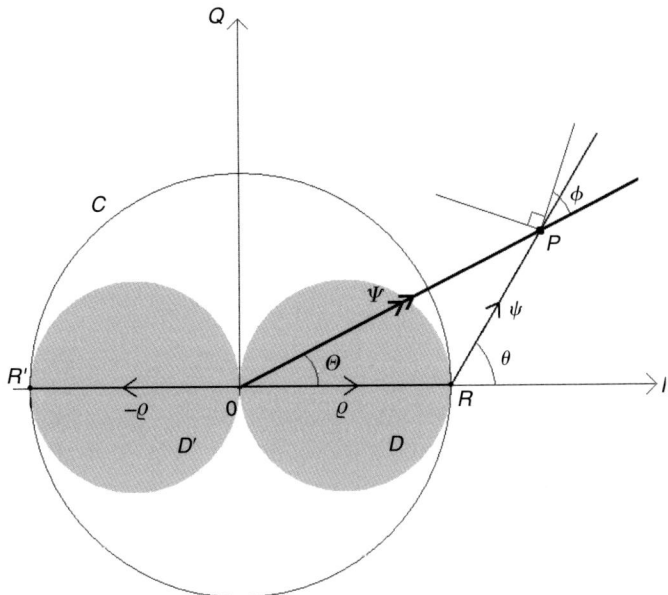

FIG. 9.1. Geometry of fluctuations for weak scattering processes depicting orthogonal dyad with respect to which resultant amplitude fluctuations decorrelate.

9.3.0.7 *Rice* The situation here is straightforward since the resultant amplitude dynamics is identical to that of Rayleigh scattering. Thus, as we have seen in Proposition 9.1, the cross term $\Sigma^{(Z,\Theta)}$ vanishes, so that $W_t^{(Z)}$, $W_t^{(\Theta)}$ are independent. The error surface S at P of Fig. 9.1 is circular, i.e. the amplitude fluctuations are isotropic and ϕ_t can take any value (both the numerator and denominator in (9.31) are identically zero).

9.3.0.8 *Homodyned K* Using Lemma 9.5 and (9.19) we find

$$\tan \phi_t = \frac{a \sin \Theta_t}{Z_t^{1/2} - a \cos \Theta_t} \tag{9.35}$$

and minus its reciprocal for the perpendicular $\psi \mapsto \phi + \pi/2$. Thus, in terms of the geometry of Fig. 9.1 (e.g. by drawing a perpendicular from R to the line OP) we see that $\phi_t = \theta_t - \Theta_t$ ($O\hat{P}R$), i.e. the (major) axis of S^{HK} coincides with that of the underlying K-scattering process, as anticipated from the differential of (9.3). The radial and angular components of the resultant amplitude fluctuations decorrelate (the diffusion coefficient $\Sigma^{(Z,\Theta)}$ of (9.19) vanishes) if the major/minor axis of the error ellipse of the K-amplitude fluctuations is aligned with the resultant amplitude (respectively the first/second factor in $\Sigma^{(Z,\Theta)}$ vanishes). In the latter case P lies on the boundary ∂D and OP, PR are perpendicular. Inside the domain D the error surface S rotates (anticlockwise for P shown in the upper half plane in Fig. 9.1) and $\Sigma^{(Z,\Theta)} <, > 0$ according as P lies in the upper/lower half plane, while the opposite situation holds for the complement $\bar{D}$.

9.3.0.9 *Generalized K* Perhaps the most interesting geometrical features emerge for generalized K-scattering. In this case, the coherent offset (e.g. a fluctuating beam) $\varrho_t = ax_t$ has intrinsic fluctuations, arising from those in the scattering population (cf. in Fig. 9.1 the boundaries ∂D and $\partial D'$ fluctuate in time). Using Lemma 9.5, Proposition 9.6, and the homodyned result (9.35) it is immediate that, for generalized K-scattering,

$$\tan \phi_t = -\frac{ax_t \sin \Theta_t}{Z_t^{\frac{1}{2}} + ax_t \cos \Theta_t} \tag{9.36}$$

(and minus the reciprocal). The above tangent corresponds to an axis of $\mathcal{S}^{\text{GK}}$ along $R'P$ (as seen e.g. by drawing a perpendicular from R' to the continuation in Fig. 9.1 of OP). In contrast to homodyned K-scattering, the symmetry axes of the error surface $\mathcal{S}^{\text{GK}}$ of the resultant amplitude are no longer aligned to those of the underlying K-amplitude. For alignment of the axes of $\mathcal{S}^{\text{GK}}$ and $\mathcal{S}^{\text{K}}$ to occur, we require the above tangent to coincide with $\tan \phi_0 = ax \sin \Theta / (Z^{\frac{1}{2}} - ax \cos \Theta)$, or minus the reciprocal, which occurs if $Z = 0$ or $Z^{\frac{1}{2}} = ax$. In the latter case P lies on the circle C shown, consistently on which PR, PR' are perpendicular. The major axis of $\mathcal{S}^{\text{GK}}$ at a general P can be identified by writing

$$d\Psi_t = (A_t + iB_t)\hat{z}_t + o(dt^{1/2}), \tag{9.37}$$

where A_t, B_t are independent real-valued (Wiener components of) Ito differentials, and $\hat{z}_t = (\Psi_t + ax_t)/|\Psi_t + ax_t|$ corresponding to a unit vector in the axial direction $R'P$. Then we have the squared relation $d\Psi_t^2 = (A_t^2 - B_t^2)\hat{z}_t^2$. Comparing with the expression for $d\Psi_t^2$ derived from Corollary 8.9 and (9.5) we find

$$\Sigma_t^{(A)} - \Sigma_t^{(B)} = \frac{\mathcal{A}|\Psi_t + ax_t|^2}{2x_t}, \tag{9.38}$$

so $\Sigma_t^{(A)} \geq \Sigma_t^{(B)}$ with equality if and only if $\mathcal{A} = 0$ or $\Psi_t + ax_t = 0$, i.e. $P = R'$. Strict inequality implies $R'P$ is the major axis of $\mathcal{S}^{\text{GK}}$ with (circular) degeneracy otherwise. From the expression for $\Sigma_t^{(Z,\Theta)}$ in (9.23), decorrelation of the radial and angular components of Ψ_t occurs if P lies on the I axis or boundary $\partial D'$. In the latter case, $R'P$ is the major axis of $\mathcal{S}^{\text{GK}}$ and OP, PR' are perpendicular ($P \neq R'$). Inside D' in the upper/lower half plane, $\Sigma^{(Z,\Theta)} >, < 0$ and $\mathcal{S}$ rotates with a corresponding orientation.

A measure of the total uncertainty ϵ_t in the resultant amplitude Ψ_t is provided by the eigenvalue product $\Sigma_t^{(\alpha)}\Sigma_t^{(\beta)} = \det\left[\Sigma_t^{ij}\right]$. For homodyned K-scattering, Corollary 8.9 and (9.30) imply

$$\Sigma_t^{(\alpha)} = \begin{cases} \frac{1}{2}\left(\frac{\mathcal{A}}{x_t}\overbrace{(Z_t - 2aZ_t^{1/2}\cos\Theta_t + a^2)}^{z_t} + \mathcal{B}x_t\right) \\ \frac{1}{2}\mathcal{B}x_t \end{cases} \qquad (9.39)$$

so that $\epsilon_t = \frac{1}{4}\mathcal{B}(\mathcal{A}z_t + \mathcal{B}x_t^2)$. Setting $\mathcal{A} = 0$, $x_t = 1$ for Rice and using Proposition 9.6 and (9.39) for generalized K-scattering we deduce the hierarchy of increasing (in the sense of the proliferation of terms that arise) uncertainties

$$\epsilon_t = \begin{cases} \frac{1}{4}\mathcal{B}^2 & \text{(Rice)} \\ \frac{1}{4}\mathcal{B}(\mathcal{A}z_t + \mathcal{B}x_t^2) & \text{(HK)} \\ \frac{1}{4}\mathcal{B}(\mathcal{A}z_t + \mathcal{B}x_t^2) + \mathcal{A}\mathcal{B}Z_t^{1/2}ax_t\cos\Theta_t & \text{(GK)}. \end{cases} \qquad (9.40)$$

These geometrical properties of the amplitude fluctuations should provide various means for anomaly detection, through the observability of the squared volatilities (cf. Field and Tough 2003a and the discussion in Chapter 12) and their departure, for $a \neq 0$, from the pure K-scattering values.

9.4 Asymptotic behaviour

The effect of the offset in the mean amplitude $\langle\Psi\rangle \neq 0$ for $a \neq 0$ is that the resulting (asymptotic) phase distributions are non-uniform. Expressions for these distributions can be calculated for the various processes we have described. We begin by deriving the joint asymptotic probability distribution functions (p.d.f.) for the cross-section, modulus amplitude, and phase, and from these deduce the marginal p.d.f.s of these quantities by integration.

9.4.0.10 Rice Noting that ψ_t is a complex Gaussian process, we see that the familiar Rice process (Rice 1954) emerges as the model for weak scattering. If we write the amplitude and phase of the scattered field as (E, Θ) their joint distribution takes the form

$$P(E,\Theta) = \frac{E\exp\left(-\left(E^2 + a^2 - 2Ea\cos\Theta\right)\right)}{\pi}. \qquad (9.41)$$

From this we can derive the familiar result for the marginal p.d.f. of the field amplitude, the *Rice distribution*,

$$P(E) = 2E\exp\left(-\left(E^2 + a^2\right)\right)I_0\left(2Ea\right), \qquad (9.42)$$

where I_0 is the modified Bessel function of the first kind. The phase distribution associated with the Rice scattering model can be obtained from (9.41) by integration over E. The result can be expressed in a reasonable closed form in terms of the error function,

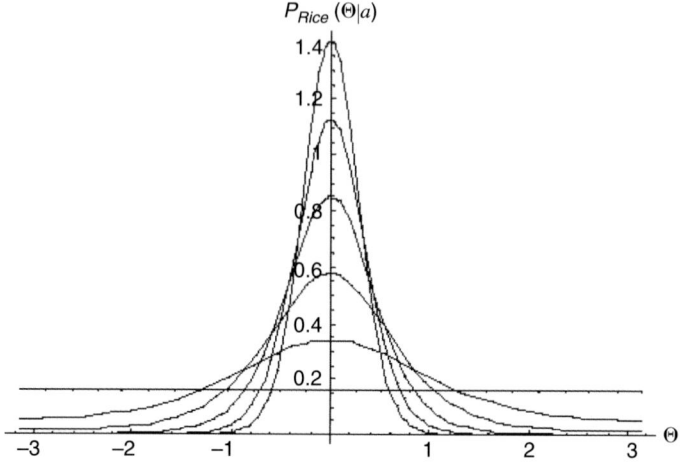

FIG. 9.2. Phase p.d.f. for the Rice scattering model, $a = 0, \frac{1}{2}, 1, \frac{3}{2}, 2, \frac{5}{2}$.

$$P(\Theta|a) = \int_0^\infty P(E, \Theta) \, dE = \frac{\exp(-a^2 \sin^2 \Theta)}{\pi} \int_0^\infty E \exp\left(-(E - a \cos \Theta)^2\right) dE$$

$$= \frac{\exp(-a^2 \sin^2 \Theta)}{\pi} \int_{-a \cos \Theta}^\infty (E + a \cos \Theta) \exp(-E^2) dE$$

$$= \frac{1}{2\pi} \exp(-a^2) + \frac{a \cos \Theta}{2\sqrt{\pi}} (1 + \mathrm{erf}(a \cos \Theta)) \exp(-a^2 \sin^2 \Theta). \quad (9.43)$$

Figure 9.2 shows the behaviour of this function, for differing values of a, whose square can be interpreted as a signal to noise power ratio.

9.4.0.11 *Homodyned K* In the case of the homodyned K-scattering process, which can be represented through (9.3), the joint p.d.f. of the cross-section, field amplitude and phase is

$$P(x, E, \Theta) = \frac{Eb^\alpha}{\pi \Gamma(\alpha)} x^{\alpha - 2} \exp(-bx) \exp\left(-(E^2 + a^2)/x\right) \exp(2Ea \cos \Theta / x). \quad (9.44)$$

The field amplitude p.d.f. associated with the homodyned K-scattering model cannot be rendered in a simple closed form for general values of the modified[25] shape parameter α. Its compound representation takes the form

---

[25] Modified in the sense that 'shape parameter' often refers to the quantity $\nu = \alpha - 1$, in a radar context.

$$P(E) = \frac{2Eb^\alpha E}{\Gamma(\alpha)} \int_0^\infty x^{\alpha-2} \exp(-bx) \exp\bigl(-(E^2+a^2)/x\bigr) I_0(2Ea/x)\,dx. \quad (9.45)$$

The asymptotic phase distribution for the homodyned K-scattering model cannot be evaluated in closed form. The compound representation of the process indicates that the phase p.d.f. can be written as

$$P(\Theta|a,b,\alpha) = \frac{b^\alpha}{\Gamma(\alpha)} \int_0^\infty P(\Theta|a,x) \exp(-bx) x^{\alpha-1}\,dx, \quad (9.46)$$

where we define

$$P(\Theta|a,x) = \frac{1}{2\pi} \exp\left(\frac{-a^2}{x}\right) + \frac{a\cos\Theta}{2\sqrt{\pi x}} \left(1 + \mathrm{erf}\left(\frac{a\cos\Theta}{\sqrt{x}}\right)\right) \exp\left(\frac{-a^2\sin^2\Theta}{x}\right). \quad (9.47)$$

This can be recast in the form

$$P(\Theta|a,b,\alpha)$$
$$= \frac{(a^2 b)^{\alpha/2}}{\pi\Gamma(\alpha)} K_\alpha\left(2\sqrt{ba^2}\right) + \frac{ab^{\frac{\alpha}{2}-\frac{1}{4}} \cos\Theta \,(a^2\sin^2\Theta)^{\frac{\alpha}{2}+\frac{1}{4}}}{\sqrt{\pi}\,\Gamma(\alpha)} K_{\alpha-1/2}\left(2\sqrt{ba^2\sin^2\Theta}\right)$$
$$+ 2\frac{a^{\alpha+1} b^{\frac{\alpha+1}{2}} \cos^2\Theta}{\pi\Gamma(\alpha)} \int_0^1 K_{\alpha-1}\left(2\sqrt{ba^2(\sin^2\Theta + t^2\cos^2\Theta)}\right)$$
$$\times (\sin^2\Theta + t^2\cos^2\Theta)^{\frac{\alpha-1}{2}}\,dt$$
$$(9.48)$$

by using the integral representation of the error function

$$\mathrm{erf}\left(a\sqrt{x}\cos\Theta\right) = \frac{2a\sqrt{x}\cos\Theta}{\sqrt{\pi}} \int_0^1 \exp\left(-s^2 a^2 x \cos^2\Theta\right)\,ds. \quad (9.49)$$

The representation (9.48) while a little arcane appears, when implemented in *Mathematica* (Wolfram 1999), to be more stable and efficient than a direct numerical integration of (9.46). The corresponding plots of the phase p.d.f. for the homodyned K-scattering model are shown in Figs. 9.3 and 9.4.

9.4.0.12 *Generalized K* To generalize the weak scattering model to the non-Gaussian regime we allow the number of steps in the biased random walk to fluctuate according to (9.4). The joint distribution of the cross-section, field amplitude, and phase now takes the form

$$P(x,E,\Theta) = \frac{Eb^\alpha}{\pi\Gamma(\alpha)} x^{\alpha-2} \exp(-bx) \exp\left(-\frac{(E^2+a^2x^2)}{x}\right) \exp(2Ea\cos\Theta). \quad (9.50)$$

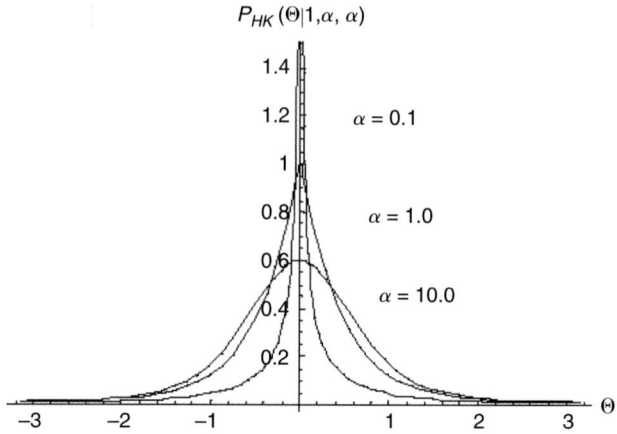

FIG. 9.3. Phase p.d.f.s derived from the homodyned K-scattering model, $\alpha = 0.1, 1.0, 10.0$.

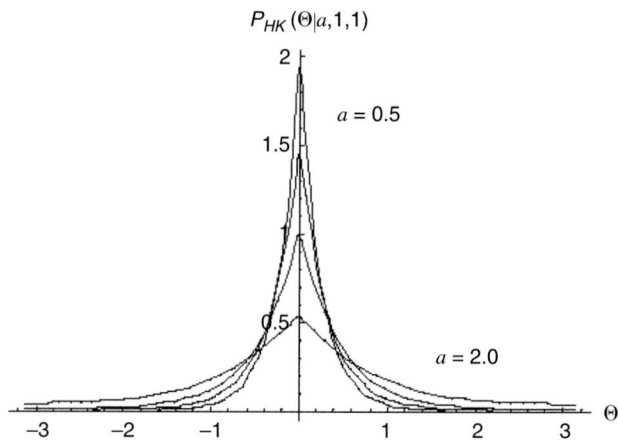

FIG. 9.4. Phase p.d.f.s derived from the homodyned K-scattering model, $a = 0.5, 1.0, 1.5, 2.0$.

(Here we have introduced the scale parameter b to relax the condition that the power in the complex Ornstein–Uhlenbeck process is taken as unity.) This provides us with the 'compound' representation of the generalized K-scattering process in accordance with (8.3). This is to be contrasted with the corresponding result for the homodyned K-scattering process above. Thus by integration we have the field amplitude p.d.f. given by

$$P(E) = \frac{4Eb^\alpha}{\Gamma(\alpha)(a^2+b)^{(\alpha-1)/2}} I_0(2Ea) K_{\alpha-1}\left(2E\sqrt{a^2+b}\right) \quad (9.51)$$

which is essentially the result obtained in Jakeman and Tough (1987) using the method of characteristic functions. The calculation of the asymptotic phase distribution for Rice scattering can be extended straightforwardly to the generalized and homodyned K-scattering models, essentially by exploiting the compound representation (8.3). Thus using (9.47) we construct

$$P(\Theta|a,b,\alpha) = \frac{b^\alpha}{\Gamma(\alpha)} \int_0^\infty P(\Theta|ax,x) \exp(-bx) x^{\alpha-1} \, dx. \quad (9.52)$$

This consists of three terms; two are straightforward while the third can be expressed in terms of a hypergeometric function. To this end we have

$$\frac{b^\alpha}{\Gamma(\alpha)} \frac{1}{2\pi} \int_0^\infty x^{\alpha-1} \exp\bigl(-(b+a^2)x\bigr) \, dx = \frac{1}{2\pi} \left(\frac{b}{b+a^2}\right)^\alpha,$$

$$\frac{b^\alpha}{\Gamma(\alpha)} \frac{1}{2\sqrt{\pi}} a \cos\Theta \int_0^\infty x^{\alpha-\frac{1}{2}} \exp\bigl(-(b+a^2\sin^2\Theta)x\bigr) \, dx \quad (9.53)$$

$$= \frac{a\cos\Theta}{\sqrt{(b+a^2\sin^2\Theta)}} \frac{\Gamma(\alpha+1/2)}{2\Gamma(\alpha)\sqrt{\pi}} \frac{1}{(1+a^2\sin^2\Theta/b)^\alpha}.$$

The third term can be evaluated by substituting (9.49) and integrating over x, thus

$$\int_0^\infty \exp\bigl(-(a^2\sin^2\Theta+b)x\bigr) x^{\alpha-1} a\sqrt{x}\cos\Theta \, \mathrm{erf}(a\sqrt{x}\cos\Theta) \, dx$$

$$= \frac{2\Gamma(\alpha+1)}{\sqrt{\pi}} a^2 \cos^2\Theta \int_0^1 \bigl(b+a^2\sin^2\Theta+a^2 s^2 \cos^2\Theta\bigr)^{-(\alpha+1)} \, ds \quad (9.54)$$

$$= \frac{2\Gamma(\alpha+1)}{\sqrt{\pi}} \frac{a^2 \cos^2\Theta}{(a^2\sin^2\Theta+b)^{\alpha+1}} {}_2F_1\left(\frac{1}{2},\alpha+1;\frac{3}{2};\frac{-a^2\cos^2\Theta}{a^2\sin^2\Theta+b}\right).$$

On bringing these results together, we obtain

$$P(\Theta|a,b,\alpha)$$
$$= \frac{1}{2\pi}\left(\frac{b}{b+a^2}\right)^\alpha + \frac{a\cos\Theta}{\sqrt{(b+a^2\sin^2\Theta)}} \frac{\Gamma(\alpha+1/2)}{2\Gamma(\alpha)\sqrt{\pi}} \frac{1}{(1+a^2\sin^2\Theta/b)^\alpha}$$
$$+ \frac{\alpha}{\pi} \frac{a^2\cos^2\Theta}{(a^2\sin^2\Theta+b)} \frac{1}{(1+a^2\sin^2\Theta/b)^\alpha} {}_2F_1\left(\frac{1}{2},\alpha+1;\frac{3}{2};\frac{-a^2\cos^2\Theta}{(a^2\sin^2\Theta+b)}\right). \quad (9.55)$$

It is interesting to compare this result with that derived in an analysis of the performance of interferometric synthetic aperture radar [eqn (53) in Tough 1991],

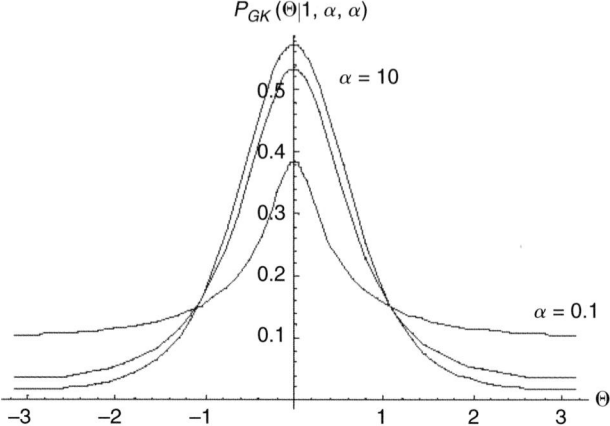

FIG. 9.5. Phase p.d.f.s derived from the generalized K-scattering model, $\alpha = 0.1, 1.0, 10.0$.

to which the above result reduces when the (modified) shape parameter α takes integer values. Figure 9.5 shows the phase p.d.f. derived from the generalized K-scattering model. We have chosen $a = 1$, $\langle x \rangle = 1$ and $\alpha = 0.1, 1, 10$. Noise with larger 'spikes', associated with lower values of α, results in a broader distribution of phase. In Fig. 9.6 we show the variation in the phase distribution with the parameter a, keeping the mean noise power $\langle x \rangle = 1$ and $\alpha = 1$. The phase distribution becomes narrower as the parameter a increases. Comparison with Fig. 9.2 shows that, while the mean noise power is the same in each, the more appreciable spikes in the character of the noise is manifest in a broader phase distribution.

The most marked difference between the phase p.d.f.s derived from the homodyned and generalized K-scattering models is evident at small values of α (i.e. less than unity), where a singular behaviour is observed at the origin. This can be seen quite clearly in Fig. 9.4. When α takes larger values, a behaviour more reminiscent of that seen in Fig. 9.2 emerges, as the noise becomes more Gaussian in character. In the case where $\alpha = 1$, the phase p.d.f. displays a cusp at the origin, irrespective of the value of a; this can be seen in Fig. 9.4. The differences between the phase p.d.f.s derived from the homodyned and generalized models can be understood qualitatively in terms of the 'signal' fluctuating with x_t in the latter, but remaining constant in the former. Jakeman and Tough (1987) discuss the implications of this difference between the models in some detail, without making explicit reference to the asymptotic phase p.d.f.s.

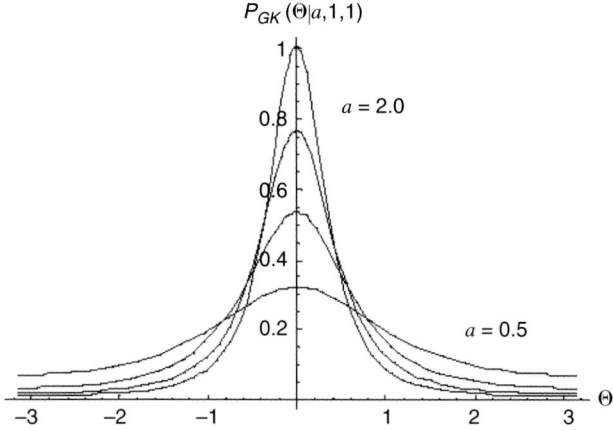

FIG. 9.6. Phase p.d.f.s derived from the generalized K-scattering model, $a = 0.5, 1.0, 1.5, 2.0$.

9.4.1 *Detailed balance*

The detailed balance condition holds (asymptotically) for each of the weak scattering processes we have described. This result follows essentially from the condition being satisfied in the case of pure K-scattering, and the ways in which the weak scattering processes can be represented as functions of an underlying K-scattering process. To complete the analysis we shall need the following result.

Lemma 9.7 *If a (complex-valued/n-dimensional) Ito diffusion process x_t^i satisfies detailed balance (at time t) then so does the transformed process $x_t^{\hat{i}} = \mathbf{f}(x_t^i)$, i.e. (in contrast to the drift) if x_t^i has vanishing current then so does $x_t^{\hat{i}}$.*

Proof Applying Ito's formula to the components $x^{\hat{i}}$ we find

$$b^{\hat{i}} = P_i^{\hat{i}} b^i + \frac{1}{2} P_{ij}^{\hat{i}} \Sigma^{ij}, \quad \Sigma^{\hat{i}\hat{j}} = P_i^{\hat{i}} P_j^{\hat{j}} \Sigma^{ij}, \tag{9.56}$$

where $P_i^{\hat{i}}$ denotes the transition matrix of partial derivatives $\partial x^{\hat{i}}/\partial x^i$, with a corresponding notation for second derivatives. Attention should be paid to the non-tensorial nature of the second term in the drift transformation, which is characteristic of the Ito calculus. The probability density transforms as $\hat{\rho} J = \rho$ where J is the Jacobian of P, i.e. $\varepsilon^{\hat{i}_1 \hat{i}_2 \ldots \hat{i}_n} J = \varepsilon^{i_1 i_2 \ldots i_n} P_{i_1}^{\hat{i}_1} P_{i_2}^{\hat{i}_2} \ldots P_{i_n}^{\hat{i}_n}$. Using the identity $\partial_j \log J \equiv P_j^k P_{jk}^{\hat{j}}$ and the relation (9.11) we deduce the vector current transformation $\mathcal{V}^{\hat{i}} = P_i^{\hat{i}} \mathcal{V}^i$, i.e. the current transforms as a tensor. The equivalent holomorphic/anti-holomorphic representation follows, via a complex change of coordinates. (The reader should compare *orig.* Field (2003) and Section 3.2 for an account of the geometrical structure of the transformations in drift and current involved here.) □

It is known that the K-scattering amplitude ψ_t (asymptotically) satisfies detailed balance, as evident from the compound representation (8.3) and the equilibrium condition (see Section 8.3.2 and also Field and Tough 2003b). As a consequence of Lemma 9.7, the relations (9.2), (9.3), (9.5), and the independence of the processes x_t, γ_t we obtain the following result.

Corollary 9.8 *The vector scattering processes (x_t, Z_t, Θ_t), in each case of weak scattering described here, namely Rice, homodyned K, and generalized K, satisfy the condition of detailed balance, asymptotically.*

Alternatively, and in rather more pedestrian fashion, detailed balance can be shown by explicit calculation using (9.11) and the expressions for the asymptotic distributions given above. In the presence of Doppler (Field and Tough 2003b) detailed balance is no longer satisfied, and the current $\mathcal{V}$ amounts to a (rigid) rotation of the Argand Ψ-plane, at a corresponding angular frequency.

We conclude the chapter with some general remarks. The K-distribution provides a useful model of the non-Gaussian statistics of strongly scattered radiation with a uniform distribution of phase. In recent work (Field and Tough 2003a,b) a description of the K-scattering process in terms of SDEs has been developed that makes direct contact with a simple underlying random walk model of scattering. In this chapter we have extended this analysis to models of weak scattering, in which the distribution of phase is non-uniform.

The K-scattering process can be derived from an isotropic random walk with a fluctuating number of steps. To incorporate a non-uniform distribution of phase, we consider a random walk on which a preferred direction or bias has been imposed. In the case where the random walk has a large, but fixed, number of steps, the Rice process emerges as a model for weak scattering. We have analysed the phase distribution associated with this model, and established the connection between its random walk formulation and a description in terms of SDEs.

The extension of the Rice scattering model to the non-Gaussian regime is effected when we allow the number of steps in the biased random walk to fluctuate. We have shown how this leads to the generalized K-scattering process discussed in Jakeman and Tough (1987) and have made contact between this model and the homodyned K-scattering process. In each case we have characterized the associated distribution of phase in detail, and have developed a description in terms of SDEs and their equivalent FPEs. This complements the earlier dynamical description of K-scattering (Field and Tough 2003a,b).

The results of this chapter have implications for detection schemes where the signal behaviour (represented by the coherent offset in the resultant amplitude) can, to a reasonable extent, be modelled in the context of ambient K-distributed noise (cf. the results reported in §4 of Field and Tough 2003a). The results of Section 9.3 indicate a method for anomaly detection based on departures in the geometry of the resultant amplitude fluctuations from that expected in the pure K-scattering case. The results should find application in adaptive imaging problems, in the de-noising of optical images (signal separation from noise, i.e. extraction of ϱ_t from Ψ_t) and anomaly detection in radar back-scatter where a (coherent) reflection contribution is involved (cf. Jakeman and Tough 1987).

10

SCATTERING FROM GENERAL POPULATIONS

In this chapter we derive the stochastic dynamics for scattering of a wavelike field from a large population of scatterers whose dynamics is arbitrary. This leads to a result concerning the observability of the scattering cross-section in terms of the resultant phase fluctuations that is independent of the population dynamics. An emergent concept is a certain notion of an ideal filter. The diffusion-based model of K-scattering arises, encountered in Chapter 8, as a special case. The experimental implications of the results in a variety of contexts are discussed later in Part III.

Motivated by the possible application of the recent results on electromagnetic scattering from random media to more general situations (e.g. medical imaging, wireless communications) than those encompassed by the K-distribution model (Field and Tough 2003a,b), the current chapter focuses on a significant new result in connection with inference of the scattering cross-section or population, in local time. A special case of the result was reported previously in Field and Tough (2003b) [see paragraph above eqn (2.35) therein], stating that the instantaneous values of the cross-section are deducible through the phase fluctuations in the scattered field. This was demonstrated theoretically in the context of K-scattering. Intriguingly, the same result holds for an arbitrary population. More precisely, given the structure of a random walk model, component phase, and step number fluctuations, the result holds for an arbitrary specification of population dynamics. In this sense, the result is a geometrical feature of (the dynamical extension of) Jakeman's random walk model with step number fluctuations (Jakeman 1980), and as such should apply to a large number of experimental situations involving interference effects of wavelike fields arising from random populations. From a filtering point of view, the result represents an improvement on Kalman/particle filtering methods, since an *exact* expression for the 'hidden' state (the population level) in terms of the additional phase degrees of freedom can be derived.

10.1 Extended random walk model

It is well known that Rayleigh scattering can be described by a random walk model for the scattered field amplitude (cf. Jakeman 1980; Tough 1987; Jakeman and Tough 1988) with a fixed number of steps. The pertinent expression for the resultant field amplitude according to this model is

$$\mathcal{E}_t^{(N)} = \sum_{j=1}^{N} a_j \overbrace{\exp\left[i\varphi_t^{(j)}\right]}^{s^{(j)}} \qquad (10.1)$$

with (constant) population size N, random phasor step $s^{(j)}$. Notice here, for the purpose of generality, the inclusion of the component amplitude 'form factors' a_j, i.e. the strength of the individual scattering components may vary throughout the medium. Also in this chapter we shall allow for the possibility that the component phases $\varphi^{(j)}$ are correlated for short times, as may occur by virtue of some special initial condition. As such, we refer to this random walk construction as the 'extended random walk model' by comparison with the simpler model of the previous chapter.

In the aforementioned references, all components of this expression are in effect considered at a given instant of time, thus not addressing the question of continuous time evolution properties or 'dynamics'. This extra structure is supplied by a (phase) diffusion model (Field and Tough 2003b) which takes the component phases $\{\varphi_t^{(j)}\}$ to be a collection of (displaced) Wiener processes evolving on a suitable timescale. Thus $\varphi_t^{(j)} = \Delta^{(j)} + \mathcal{B}^{\frac{1}{2}} W_t^{(j)}$, with component random initialization $\Delta^{(j)}$ uniformly distributed on the interval $[0, 2\pi)$. In situations where the component phasors $s^{(j)}$ are aligned initially (e.g. for the received '$T2$' signal arising in magnetic resonance imaging, described e.g. in Ernst et al. 1987 and discussed in Chapter 14) (cf. also Field and Bain 2008), the initializations $\Delta^{(j)}$ are identical for all j, whereas in some cases (e.g. the statistical description of radar scattering from the sea surface, cf. §4(b) in Field and Tough 2003a) it is more appropriate to draw $\Delta^{(j)}$ independently. In any case, these primitive assumptions enable us to derive the dynamics of Rayleigh scattering, essentially from first principles. The stochastic differential of (10.1), according to Ito's formula (e.g. Oksendal 1998; Karatzas and Shreve 1988), is given by

$$d\mathcal{E}_t^{(N)} = \sum_{j=1}^{N} a_j \left(i d\varphi_t^{(j)} - \frac{1}{2} d\varphi_t^{(j)2}\right) \exp\left[i\varphi_t^{(j)}\right]. \qquad (10.2)$$

If we write $d\zeta_t$ for the first term on the right hand side above, then for $t \geq T$, where T is the 'phase decoherence' time such that $\{\varphi_t^{(j)} \mid t \geq T\}$ have negligible correlation, we have $|d\zeta_t|^2 = \left(\sum_j a_j^2\right) \mathcal{B} dt$, and therefore $d\zeta_t = \left(\sum_j a_j^2\right)^{1/2} \mathcal{B}^{1/2} d\xi_t$ where ξ_t is a complex-valued Wiener process (satisfying $|d\xi_t|^2 = dt$, $d\xi_t^2 = 0$). Defining the (normalized) Rayleigh amplitude by $\gamma_t = \lim_{N \to \infty} \left[\mathcal{E}_t^{(N)}/N^{\frac{1}{2}}\right]$ leads to the resultant dynamics (cf. Field and Tough 2003b).

Proposition 10.1 *For sufficiently large times $t \geq T$ the dynamics of Rayleigh scattering is given by the complex Ornstein–Uhlenbeck equation*

$$d\gamma_t = -\frac{1}{2}\mathcal{B}\gamma_t dt + \mathcal{B}^{\frac{1}{2}}\langle a^2\rangle^{\frac{1}{2}} d\xi_t. \tag{10.3}$$

If $\{\Delta^{(j)}\}$ are assumed independent then the result holds for arbitrarily small times.

Remarks. *On the asymptotic distribution of phase.* If the phases are initialized at some value, say the real direction in the complex plane, then the distribution of the resultant Rayleigh amplitude is Gaussian, of non-zero mean, by virtue of the central limit theorem applied to the i.i.d. collection of random phasors (with a square root scaling). Observe that, for small times, the major axis of its error surface $\mathcal{S}$ – an ellipse – is oriented in the direction of the *imaginary* axis, since the magnitude of the fluctuations (for each component phasor) in that direction is dominant. As relaxation occurs, i.e. the asymptotic equilibrium distribution is approached, the surface $\mathcal{S}$ stretches in the real direction, and tends to a circular geometry in the limit $t \to \infty$. At the same time, the mean value – geometrically the centre of gravity of the ellipse – tends to the origin of the complex plane. Accordingly, the resultant asymptotic distribution of phase is uniform.

If we re-scale the (Rayleigh) amplitude according to $\gamma_t \mapsto \langle a^2 \rangle^{-1/2}\gamma_t$, then the re-scaled field satisfies (10.3) with the form factors equal to unity. In what follows we shall therefore assume the field to be scaled in this way, i.e. $\langle a^2 \rangle = 1$. In the case of a fluctuating number of steps $N \mapsto N_t$ in (10.1), we define the (continuous-valued) cross-section as $x_t = \lim_{N_t \to \infty} [N_t/\bar{N}]$. The resultant (normalized) amplitude $\psi_t = \lim_{N \to \infty} [\mathcal{E}^{(N_t)}/\bar{N}]$ therefore has the compound representation

$$\psi_t = x_t^{\frac{1}{2}}\gamma_t, \tag{10.4}$$

where $\gamma_t = \lim_{N \to \infty}\left[\mathcal{E}_t^{(N_t)}/N_t^{\frac{1}{2}}\right]$, and in which x_t and γ_t are independent processes. The intensity z_t has the compound representation $z_t = x_t u_t$, where $u_t = |\gamma_t|^2$ is the instantaneous intensity of the component (unit power) Rayleigh process (cf. the analysis of asymptotic behaviour and propagators in §s III, IV of Field and Tough 2003b).

It is worth clarifying at this point the precise meaning and definitions of the various amplitudes that have occurred in the exposition of the random walk model. We begin with $\mathcal{E}$, as the superposition of N random phasors, where N is fixed. For a large population, $N \to \infty$, and the root mean square (r.m.s.) of $\mathcal{E}$ tends to infinity. Thus, to obtain a finite resultant in the limit of an asymptotically large number of scatterers, we define a 'normalized' Rayleigh amplitude γ, by dividing through by the r.m.s. value $N^{\frac{1}{2}}$. (Equivalently, we could absorb this normalization into the form factors a_j.) The term 'Rayleigh' refers to the fact that the number of scatterers is fixed. In the general case that the scattering population fluctuates in time, we define the 'normalized' amplitude ψ as in the case of Rayleigh scattering, dividing $\mathcal{E}$ by the r.m.s. value $\bar{N}^{\frac{1}{2}}$, where now the number of terms N_t in the random walk fluctuates. Re-arranging the resulting expression produces the compound representation of the resultant amplitude (10.4), for a general scattering process. The Rayleigh amplitude is then recovered if the cross-section is unity.

A corresponding dynamical situation for 'weak' scattering processes, i.e. where the field ψ_t lies in (weak) superposition with a coherent offset signal ϱ_t, should be possible for a general type population as a generalization of the results described in Chapter 9 (*orig.* Field and Tough 2005), although this extension is not addressed in the current monograph.

10.2 Generalized dynamics

In this section we propose a general scheme for describing scattering/interference of wavelike fields from random media, for which the statistical characteristics of the (scattering) population are taken to be arbitrary. With regard to the cross-section and intensity variables, in 'Bayesian' terms one may write $\mathbf{P}(x|z) \propto \mathbf{P}(z|x)\mathbf{P}(x)$ and interpret $\mathbf{P}(z|x)$ as the 'likelihood' function L, $\mathbf{P}(x)$ as the 'prior' P and $\mathbf{P}(x|z)$ as the 'posterior' distribution. In K-scattering, the constant (with respect to x) of proportionality is the reciprocal K-distribution for the intensity z. Our development shall entail the following, that we first preserve the likelihood L as the Rayleigh distribution, cf. the universality (under appropriate conditions) of arguments of central limit theorem type, second modify the prior P, i.e. consider general (endogenous) population dynamics appropriate to more complex population processes, under the assumption that x_t remains an Ito process (cf. Field and Tough 2003a), and third preserve the mathematical structure of the random walk model (10.1) describing the resultant amplitude process.

Accordingly, we specify that the underlying 'signal' x_t is an Ito process that satisfies the generalized (in the sense of its relationship with the K-scattering model) stochastic differential equation (SDE)

$$\mathrm{d}x_t = \mathcal{A}b_t\mathrm{d}t + (2\mathcal{A}\Sigma_t)^{\frac{1}{2}}\mathrm{d}W_t^{(x)} \tag{10.5}$$

in which the drift and diffusion parameters b_t, Σ_t are, respectively, (real-valued) stochastic processes, not necessarily Ito processes, adapted to the filtration $\mathcal{F}_t^{(x)}$ corresponding to the Wiener process $W_t^{(x)}$. In other words, the continuous population dynamics is taken to lie within the general category of Ito processes. The special case of a diffusive population behaviour arises when the SDE parameters are functions of state, i.e. $b_t = b(t, x_t)$ and $\Sigma_t = \Sigma(t, x_t)$ for given functions $b(\cdot, \cdot)$ and $\Sigma(\cdot, \cdot)$, in which case a corresponding Fokker–Planck description for the time evolution of the probability density is possible (e.g. Risken 1989). The case of K-scattering (a special type of diffusion model) is obtained by setting $b(t, x) = (\alpha - x)$, $\Sigma(t, x) = x$, and arises as the continuous-valued (large N) limit of the birth–death–immigration (BDI) model (see Bartlett 1966; Field and Tough 2003b). We shall not require that x_t be a diffusion in what follows, however. The generalized dynamics of the resultant amplitude process can now be derived according to the scheme outlined at the end of the previous section. For arbitrary γ_t, x_t an application of Ito's formula to (10.4) yields

$$\frac{\mathrm{d}\psi_t}{\psi_t} = \frac{\mathrm{d}\gamma_t}{\gamma_t} + \frac{\mathrm{d}x_t}{2x_t} - \frac{\mathrm{d}x_t^2}{8x_t^2}. \tag{10.6}$$

This enables the resultant amplitude dynamics to be calculated under the assumption that γ_t is a unit power Rayleigh process according to (10.3), with unit form factors.

Proposition 10.2 *The generalized resultant amplitude dynamics is given by*

$$\frac{\mathrm{d}\psi_t}{\psi_t} = \left[\mathcal{A}\left(\frac{b_t}{2x_t} - \frac{\Sigma_t}{4x_t^2}\right) - \frac{1}{2}\mathcal{B}\right]\mathrm{d}t + \left(\frac{\mathcal{A}\Sigma_t}{2x_t^2}\right)^{\frac{1}{2}}\mathrm{d}W_t^{(x)} + \left(\frac{\mathcal{B}^{\frac{1}{2}}}{\gamma_t}\right)\mathrm{d}\xi_t. \tag{10.7}$$

Observe that $\partial/\partial\mathcal{B}$ acting on the drift/volatility parameters in (10.7) yields expressions that are independent of b_t, Σ_t, as expected from the endogenous specification of population dynamics (10.5). Using the vanishing of the Ito products $\mathrm{d}\xi_t^2$, $\mathrm{d}\xi_t \mathrm{d}W_t^{(x)}$, and the property $|\mathrm{d}\xi_t|^2 = \mathrm{d}t$, the above result yields the squared amplitude fluctuations as follows.

Corollary 10.3

$$\left(\frac{\mathrm{d}\psi_t}{\psi_t}\right)^2 = \frac{\mathcal{A}\Sigma_t}{2x_t^2}\mathrm{d}t, \tag{10.8}$$

$$|\mathrm{d}\psi_t|^2 = \left(\frac{\mathcal{A}\Sigma_t z_t}{2x_t^2} + \mathcal{B}x_t\right)\mathrm{d}t. \tag{10.9}$$

The generalized intensity dynamics can be computed from Proposition 10.2 and the identity $\mathrm{d}z_t \equiv \psi_t \mathrm{d}\psi_t^* + \psi_t^* \mathrm{d}\psi_t + \mathrm{d}\psi_t \mathrm{d}\psi_t^*$.

Proposition 10.4 *The generalized intensity SDE is given by*

$$\mathrm{d}z_t = \left[\mathcal{A}\left(\frac{b_t z_t}{x_t}\right) + \mathcal{B}(x_t - z_t)\right]\mathrm{d}t + (2\mathcal{A}\Sigma_t)^{\frac{1}{2}}\left(\frac{z_t}{x_t}\right)\mathrm{d}W_t^{(x)} + (2\mathcal{B}x_t z_t)^{\frac{1}{2}}\mathrm{d}W_t^{(r)} \tag{10.10}$$

where

$$(\gamma_t^* \mathrm{d}\xi_t + \gamma_t \mathrm{d}\xi_t^*) \equiv \left(\frac{2z_t}{x_t}\right)^{\frac{1}{2}}\mathrm{d}W_t^{(r)}. \tag{10.11}$$

The intensity squared volatility is

$$\mathrm{d}z_t^2 = 2z_t\left(\frac{\mathcal{A}\Sigma_t z_t}{x_t^2} + \mathcal{B}x_t\right)\mathrm{d}t. \tag{10.12}$$

The propositions above reduce to the appropriate expressions in K-scattering (cf. Field and Tough 2003a,b) for appropriate choice of b, Σ. In respect of the generalized resultant phase dynamics, recall (Field and Tough 2003b) the identity for the phase differential in terms of the amplitude

$$\mathrm{d}\theta_t \equiv \Im\left[\frac{\mathrm{d}\psi_t}{\psi_t} - \frac{1}{2}\left(\frac{\mathrm{d}\psi_t}{\psi_t}\right)^2\right], \qquad (10.13)$$

where $\Im$ denotes the imaginary part. Since the right-hand side of (10.8) is real-valued, only the first term on the right-hand side of (10.13) contributes to $\mathrm{d}\theta_t$, in respect of which

$$\Im\left[\frac{\mathrm{d}\psi_t}{\psi_t}\right] = \frac{\mathcal{B}^{\frac{1}{2}}}{2i}\left(\frac{\mathrm{d}\xi_t}{\gamma_t} - \frac{\mathrm{d}\xi_t^*}{\gamma_t^*}\right). \qquad (10.14)$$

Thus we can deduce the phase behaviour for a general population.

Proposition 10.5 *The generalized resultant phase dynamics is given by the SDE*

$$\mathrm{d}\theta_t = \left(\frac{\mathcal{B}x_t}{2z_t}\right)^{\frac{1}{2}} \mathrm{d}W_t^{(\theta)} \qquad (10.15)$$

where $W_t^{(\theta)}$ satisfies

$$\frac{1}{i}(\gamma_t^*\mathrm{d}\xi_t - \gamma_t\mathrm{d}\xi_t^*) \equiv \left(\frac{2z_t}{x_t}\right)^{\frac{1}{2}} \mathrm{d}W_t^{(\theta)}. \qquad (10.16)$$

Observe that, in contrast to the situation for the resultant amplitude and intensity SDEs (10.7) and (10.10), this is *functionally* identical to the corresponding result in K-scattering (i.e. independent of the population parameters b_t, Σ_t), the essential difference lying in the *evolutionary* structure of the processes x_t, z_t. Observe from (10.11), (10.16) that the radial and angular fluctuations in the resultant amplitude are statistically independent in the general case. The squared phase volatility obtained from (10.15) leads to the central result of the chapter.

Theorem 10.6 *The instantaneous values of the scattering cross-section are observable through the intensity-weighted squared phase fluctuations according to*

$$x_t = \frac{2}{\mathcal{B}}z_t\frac{\mathrm{d}\theta_t^2}{\mathrm{d}t} \qquad (10.17)$$

if x_t is an Ito process, not necessarily a diffusion, and throughout space and time.

The (experimental) significance of this result is that the relation (10.17) is exact and moreover independent of the dynamics of x_t. The result resembles (but is distinct from) the minimal variance of the intensity-weighted phase derivative

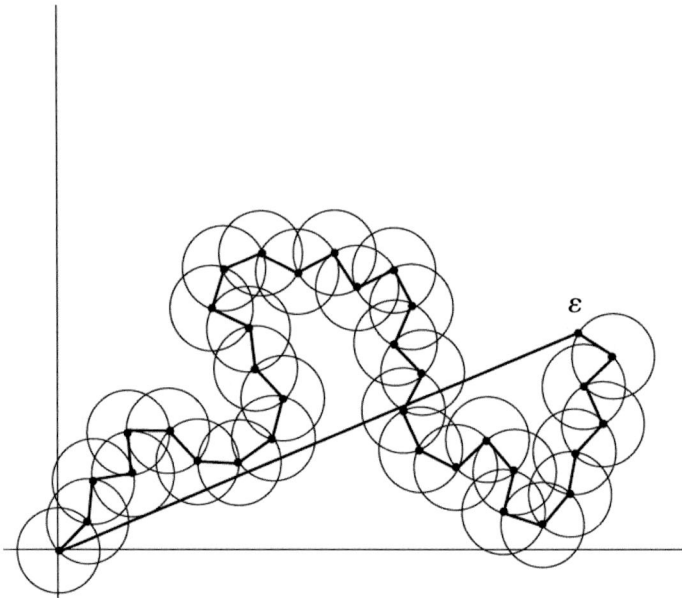

FIG. 10.1. Geometry of random walk for generalized scattering process – generically (a.e.) each component phasor point lies on two circles.

discussed in Jakeman *et al.* (2001), for differentiable processes. In the present situation however, the processes considered are not differentiable, and instead the *squared* phase differential arises. Since the elements of the random walk model (10.1), preserved by the generalized framework given in the exposition of this section for dealing with general populations, are the only essential ingredients involved here, the result is geometrical in nature (cf. Fig. 10.1). With regard to this geometry, the derivations above show that the dynamics of the resultant field are not affected if the radii for each component phasor are drawn independently from an arbitrary probability distribution. (The result of the theorem was anticipated from a physical point of view previously in Field and Tough 2003*a*; see discussion following Prop. 4.1. therein.)

A slight complication is posed in the computation of $d\theta_t^2$ from experimental data, owing to the discontinuous-valued behaviour of θ_t at coordinate intervals of 2π. This is resolved by instead using the (continuous-valued) 'phase-wrapped' process $w_t = \exp(i\theta_t)$, whose stochastic differential is $dw_t = \exp(i\theta_t)[id\theta_t - \frac{1}{2}d\theta_t^2]$, which enables the squared phase fluctuations to be computed from the single-valued process w_t via $|dw_t|^2 = d\theta_t^2$. In respect of discrete-time implementation, we remark that if W_t is a Wiener process, then $\delta W_t = W_{t+h} - W_t$ is normally distributed as $\mathcal{N}(0, h)$, so that its square is a 'chi-squared' $\chi^2(1)$ variable. The sum of n such variables is therefore distributed as $\chi^2(n)$, from which an estimate of dq_t^2 from δq_t can be obtained (via the weak law of large numbers) by considering the interval from t to $t + \delta t$ divided into n 'pulse' intervals each of length h and letting $n \to \infty$ before taking the limit $\delta t \to \mathrm{d}t$ (Howison 2004; see also Higham 2004 and the exposition to §4 in Field and Tough 2003a). In this respect we observe the following significant result.

Proposition 10.7 *In order to achieve an improved signal to noise ratio (SNR), it is sufficient merely to increase the pulse-rate n, without necessarily requiring a high amplitude signal x_t.*

The implications of this idea are discussed in Chapter 14 in an experimental context, and illustrated with some synthetically generated data. The reader should compare the remarks in Section 14.1 for issues concerning discrete sampling and bounded pulse rate, and the optimal choice of a smoothing parameter that effects the result of the theorem above. The structure of (10.5), (10.10) could be interpreted as an instance of the 'generalized' Kalman filter (see Ch. 6 in Oksendal 1998) in which the unknown state x_t is to be estimated from observations of z_t. It is significant that in this situation the dynamics of the filter stem from first principles and that the resulting statistics are non-Gaussian (notwithstanding the Gaussian nature of the Wiener process). The noise originates through two components, namely the 'intrinsic' system noise $W_t^{(x)}$ which derives from fluctuations in the (endogenously specified) population model, and the 'measurement' noise ξ_t arising from the particulars of the wavelike interference effects. The latter should be viewed as an exogenous 'device' whose purpose is to probe the true underlying state of the system that is of primary interest, in this case the 'signal' x_t.

Our development has demonstrated (an instance of) how, instead of attempting to filter the received signal to eliminate the noise (e.g. via a Kalman or particle filter), one can exploit the statistical fluctuation properties of the noise to infer the *exact* values of the underlying signal. This notion might appropriately be termed an *ideal filter* and constitutes a shift of viewpoint from the various conventional approaches to enhancement of signal to noise. We discuss the experimental implications of this framework in greater detail in Chapter 14.

Finally, we remark in proof that stochastic models of the kind discussed in this part have recently been applied successfully to modelling of the wireless channel in communications, in which the latter can be viewed as a multi-path propagation problem consisting of a combination of phase fluctuations and Doppler effects for each component path. The theoretical results predict modifications to the channel spectra that are observed experimentally, and a corresponding state-space model can be constructed (Feng et al. 2007; Feng and Field 2008).

Part III

Simulation and experiment

In this part we develop simulation techniques for scattering processes and describe some experimental tests that verify our theoretical development. This develops the interrelationships between theoretical prediction, real, and simulated data, and is illustrated in a specific case of scattering from a general population.

Chapter 11 provides a detailed account of the simulation of K-scattering processes, including the influence of Doppler, target returns, and improved accuracy second-order algorithms for numerical integration of the gamma process. In Chapter 12 we provide two independent experimental tests of the proposed stochastic theory, and provide a scheme for anomaly detection. The first in Section 12.1 is a study of data gathered from electromagnetic scattering from a random phase screen, in the absence of anomalies. A study is made of the (forward) scattered intensity which is observed to be very closely K-distributed. The volatility function of the intensity is measured and observed to correlate very strongly with the instantaneous square root intensity, thus providing a calibration of the stochastic model. This calibration setting is applied to an independent set of (complex-valued) radar return data in Section 12.2. The radar scattering is from a region of the ocean surface in which a tethered anomalous reflecting body is placed and whose location is known. It is previously known that such data typically contains a large number of spurious 'clutter spike' anomalies and that the phase behaviour is an important indicator of a genuine anomaly (Luttrell 2001). Accordingly, we derive a complex-valued stochastic differential equation (SDE) for the amplitude of the scattered radiation, under the assumption that the phase behaviour is uniform. We exhibit a strong correlation between the observed versus the theoretical prediction of the volatility for the complex-valued process in the clutter domain. At the location of the anomaly this correlation becomes weak, which provides a means of detection, isolating the anomaly to a high degree of accuracy. With the theory firmly established from these experiments, Chapter 13 provides a discussion of the non-linear character of the stochastic dynamics of radar sea clutter, from independent experimental and theoretical points of view.

The part concludes in Chapter 14 with some simulations that illustrate the concepts of scattering from general populations as developed earlier in Chapter 10. The implications of these results for tracking populations in various physical contexts are discussed.

11

SIMULATION OF K-SCATTERING

We have seen in Chapter 8 that the stochastic differential equations (SDEs) arising in the description of scattering generically contain Wiener fluctuating terms whose noise power is a function of the state of the system, i.e. the system is said to have 'multiplicative noise' or, in other words, it falls under the category of a 'stochastic volatility' model. In this chapter we develop the tools necessary for the numerical integration/simulation of such systems in a fairly complete general context. From this analysis the case of additive noise, for which the noise power is independent of the state of the system, will emerge as a special case. This framework is applied to Rayleigh and gamma processes that are of special interest in K-scattering. We examine in detail the effect of Doppler for discrete time-series data and describe the simulation of a target in a clutter environment. The chapter ends with an investigation of the refinement in terms of second-order algorithms for the gamma process.[26]

11.1 Iterative schemes

Consider a differential equation expressed in the Langevin form

$$\frac{\mathrm{d}x}{\mathrm{d}t} = F(x) + \sigma(x) f(t) \qquad (11.1)$$

in which $f(t)$ is a white noise process, whose integral is the Wiener process. The interpretation of this equation has, in the statistical physics literature, been the source of considerable controversy and dispute. It is no longer possible to associate a Langevin equation of this form unambiguously with a Fokker–Planck equation (FPE), without additional specification of how the fluctuating terms undergo integration.[27] Pusey and Tough (1982) have discussed the iterative solution of equations of this type in some detail, in such a way that the ambiguities are avoided. Indeed in the mathematical and engineering literature, the utility of such equations has long been recognized, and considerable effort has been expended in accommodating them within a consistent and rigourous calculus.

The analysis given by Pusey and Tough *ibid.* produces results that are in accordance with this Ito, as opposed to Stratonovich, calculus, which is the convention followed throughout this book (cf. Appendix B). The resulting theory

---

[26]The author acknowledges the contribution of Dr. Robert Tough with regard to the majority of material in this chapter.

[27]Indeed some workers, such as Van Kampen (1981), have characterized such equations as being devoid of meaning.

of SDEs has found widespread application in signal processing, control theory and mathematical finance (e.g. Oksendal 1998).

$$dx_t = F(x_t)\,dt + \sigma(x_t)\,dW_t \qquad (11.2)$$

which is to be understood in the Ito interpretation, as developed in Chapter 2. We now distinguish between integration with respect to time and with respect to the Brownian measure and write the associated integral equation in the form

$$\Delta x_t = x_t - x_0 = \int_0^t F(x_u)\,du + \int_0^t \sigma(x_u)\,dW_u. \qquad (11.3)$$

It is natural to develop the drift F and volatility σ in Taylor series, which yields

$$\Delta x_t = \sum_n \frac{F^{(n)}}{n!} \int_0^t (\Delta x_u)^n\,du + \sum_n \frac{\sigma^{(n)}}{n!} \int_0^t (\Delta x_u)^n\,dW_u. \qquad (11.4)$$

A solution to this is now developed via iteration. We calculate terms up to $O(t^2)$; Pusey and Tough *ibid.* carried through their analysis to this level in the multiplicative noise case. Taking only the $n = 0$ contributions gives us the following terms

$$\Delta x_t = Ft + \sigma W_t. \qquad (11.5)$$

This simple stochastic analogue of the Euler integration algorithm forms the basis of many widely used Brownian dynamics algorithms that incorporate hydrodynamic interactions (Ermak and McCammon 1978). Taking the first four terms in the Taylor expansions in (11.4) gives us

$$\Delta x_t = Ft + \sigma W_t + F^{(1)} \int_0^t \Delta x_u\,du + \sigma^{(1)} \int_0^t \Delta x_u\,dW_u + \frac{1}{2}F^{(2)} \int_0^t (\Delta x_u)^2\,du$$
$$+ \frac{1}{2}\sigma^{(2)} \int_0^t (\Delta x_u)^2\,dW_u + \frac{1}{6}F^{(3)} \int_0^t (\Delta x_u)^3\,du + \frac{1}{6}\sigma^{(3)} \int_0^t (\Delta x_u)^3\,dW_u + \cdots.$$
$$(11.6)$$

We consider the orders of the various terms generated by this process, and make use of the properties $Ft = O(t)$, $W_t = O(\sqrt{t})$. Integration with respect to u raises the order by one while integration with respect to W_u raises it by one half.

To generate terms of order two, we must therefore keep terms up to order t in the integrand of

$$F^{(1)} \int_0^t \Delta x_u\,du \qquad (11.7)$$

while terms up to $O(t^{3/2})$ must be retained in the integrand of

$$\sigma^{(1)} \int_0^t \Delta x_u \, dW_u. \tag{11.8}$$

Not all these are provided by the lowest order solution (11.5) when it is introduced into (11.1). Nonetheless, it does generate terms of orders t, $t^{3/2}$, t^2. These in turn should be re-inserted into (11.1) to produce extra terms in the iteration. Thus, from (11.5) alone, we have

$$\begin{aligned}\Delta x_t = {}& Ft + \sigma W_t + FF^{(1)}\frac{t^2}{2} \\ & +F^{(1)}\sigma \int_0^t W_u du \left(O\left(t^{3/2}\right)\right) + F\sigma^{(1)} \int_0^t u dW_u \left(O\left(t^{3/2}\right)\right) \\ & +\sigma^{(1)}\sigma \int_0^t W_u dW_u \left(O\left(t\right)\right) \\ & +\frac{1}{2}F^{(2)}\sigma \int_0^t W_u^2 du \left(O\left(t^2\right)\right) + \frac{1}{2}\sigma^{(2)}\sigma^2 \int_0^t W_u^2 dW_u \left(O\left(t^{3/2}\right)\right) \\ & +\sigma^{(2)}\sigma F \int_0^t u W_u dW_u \left(O\left(t^2\right)\right) + \frac{\sigma^{(3)}\sigma^3}{6} \int_0^t W_u^3 dW_u \left(O\left(t^2\right)\right).\end{aligned} \tag{11.9}$$

Here the order of each term in t is included in parentheses, where it is not obvious. Terms that are of higher order in t than the second have been omitted. By inspection we see that the term

$$\sigma^{(1)}\sigma \int_0^t W_u \, dW_u \tag{11.10}$$

should be incorporated into the integrand in (11.7) and in that of

$$\frac{1}{2}\sigma^{(2)} \int_0^t (\Delta x_u)^2 \, dW_u. \tag{11.11}$$

There they give rise to terms of the form

$$F^{(1)}\sigma^{(1)}\sigma \int_0^t du \int_0^u W_s \, dW_s \tag{11.12}$$

and
$$\sigma^{(2)}\sigma^{(1)}\sigma^2 \int_0^t dW_u W_u \int_0^u dW_s W_s. \tag{11.13}$$

Each of these is of $O(t^2)$ and so need be iterated no more.

We now turn our attention to (11.8); including all terms up to $O(t^{3/2})$ obtained thus far we generate the following new terms

$$\left(\sigma^{(1)}\right)^2 \sigma \int_0^t dW_u \int_0^u W_s \, dW_s + \sigma^{(1)} F^{(1)} \sigma \int_0^t dW_u \int_0^u W_s \, ds$$

$$+ \sigma^{(1)} F \sigma \int_0^t dW_u \int_0^u s \, dW_s + \frac{1}{2}\sigma^{(1)}\sigma^{(2)}\sigma^2 \int_0^t dW_u \int_0^u W_s^2 \, dW_s. \tag{11.14}$$

The last three of these terms are of order t^2; the first, however, is $O(t^{3/2})$ and so should be iterated once more. This leads us to a contribution of the form

$$\left(\sigma^{(1)}\right)^3 \sigma \int_0^t dW_u \int_0^u dW_s \int_0^s dW_v W_v. \tag{11.15}$$

This process of successive re-substitution enables us to generate all the terms of a required order in a systematic fashion; as a result we obtain a solution correct to $O(t^2)$ of the form:

$$\Delta x_t = Ft + \sigma W_t + FF^{(1)}\frac{t^2}{2} + F^{(1)}\sigma \int_0^t W_u du + F\sigma^{(1)} \int_0^t u dW_u + \sigma^{(1)}\sigma \int_0^t W_u dW_u$$

$$+ \frac{1}{2}F^{(2)}\sigma^2 \int_0^t W_u^2 du + \frac{1}{2}\sigma^{(2)}\sigma^2 \int_0^t W_u^2 dW_u + \sigma^{(2)} F\sigma \int_0^t u W_u dW_u$$

$$+ F^{(1)}\sigma^{(1)}\sigma \int_0^t du \int_0^u W_s dW_s + \sigma^{(2)}\sigma^{(1)}\sigma^2 \int_0^t dW_u W_u \int_0^u dW_s W_s$$

$$+ \frac{\sigma^{(3)}\sigma^3}{6} \int_0^t W_u^3 dW_u + \left(\sigma^{(1)}\right)^2 \sigma \int_0^t dW_u \int_0^u W_s dW_s + \sigma^{(1)} F^{(1)} \sigma \int_0^t dW_u \int_0^u W_s$$

$$+ \left(\sigma^{(1)}\right)^2 F \int_0^t dW_u \int_0^u s dW_s + \frac{1}{2}\sigma^{(1)}\sigma^{(2)}\sigma^2 \int_0^t dW_u \int_0^u W_s^2 dW_s$$

$$+ \left(\sigma^{(1)}\right)^3 \sigma \int_0^t dW_u \int_0^u dW_s \int_0^s dW_v W_v. \tag{11.16}$$

This process could be carried on to higher order; the number of terms rapidly becomes very large. Perhaps the most striking qualitative difference between this and the corresponding result obtained in the additive noise case is the presence of a term with a *linear* time dependence in the stochastic part of the increment.

Now that we have developed an iterative solution we calculate the expectation value of the increment in x occurring over the time t. All terms in (11.16) containing odd powers of W average to zero. Among those containing even powers of W the majority of the expectation values go to zero as a consequence of the Ito rule (2.6)

$$\mathbf{E}\left(\int_0^t dW_u \int_0^u dW_s g(u,s)\right) = 0. \tag{11.17}$$

For example, we consider the simple case where g is unity:

$$\int_0^t dW_u \int_0^u dW_s - \int_0^t dW_u W_u = \frac{W_t^2 - t}{2}; \quad \mathbf{E}\left(W_t^2\right) = t,$$

$$\mathbf{E}\left(\int_0^t dW_u \int_0^u dW_s\right) = 0. \tag{11.18}$$

Here we observe the occurrence of an extra term arising in the integration by parts within the Ito calculus; it is this term that ensures that the expectation value of the iterated integral is indeed zero. Examination of each of the terms occurring in (11.16) shows that

$$\mathbf{E}(\Delta x_t) = Ft + FF^{(1)}\frac{t^2}{2} + \frac{1}{2}F^{(2)}\sigma \int_0^t \mathbf{E}\left(W_u^2\right) du$$

$$= Ft + FF^{(1)}\frac{t^2}{2} + \frac{1}{4}F^{(2)}\sigma t^2. \tag{11.19}$$

This is identical with the corresponding result obtained in the additive noise case, to second order in time.

We now wish to evaluate the mean square of the stochastic part of the increment, i.e.

$$\Delta \hat{x}_t = \Delta x_t - Ft - FF^{(1)}\frac{t^2}{2} \tag{11.20}$$

again retaining only terms up to second order in t. (We have introduced the caret notation because superscript notation has already been used in this case.) Thus we need only consider the terms in (11.16) up to and including those that are $O(t^{3/2})$,

$$\Delta \hat{x}_t = \sigma W_t + F^{(1)}\sigma \int_0^t W_u \, du + F\sigma^{(1)} \int_0^t u \, dW_u$$
$$+ \sigma^{(1)}\sigma \int_0^t W_u \, dW_u + \frac{1}{2}\sigma^{(2)}\sigma \int_0^t W_u^2 \, dW_u + O\left(t^2\right). \quad (11.21)$$

From this we see that

$$\mathbf{E}\left((\Delta \hat{x}_t)^2\right) = \sigma^2 \mathbf{E}\left(W_t^2\right) + \left(\sigma^{(1)}\sigma\right)^2 \mathbf{E}\left(\left(\int_0^t W_u \, dW_u\right)^2\right)$$
$$+ 2F^{(1)}\sigma^2 \mathbf{E}\left(W_t \int_0^t W_u \, du\right)$$
$$+ 2F\sigma^{(1)}\sigma \mathbf{E}\left(W_t \int_0^t u \, dW_u\right) + \sigma^{(2)}\sigma^2 \mathbf{E}\left(W_t \int_0^t W_u^2 \, dW_u\right) + O\left(t^2\right). \quad (11.22)$$

The various expectation values can be evaluated, using the Ito calculus. We recall the familiar relation $\mathbf{E}[W_t^2] = t$. The Ito isometry (see Section 2.1.1) allows us to evaluate the next contribution:

$$\mathbf{E}\left(\left(\int_0^t W_u \, dW_u\right)^2\right) = \int_0^t \mathbf{E}\left(W_u^2\right) du = \frac{t^2}{2}. \quad (11.23)$$

The remaining averages can be obtained as follows

$$\mathbf{E}\left(W_t \int_0^t W_u \, du\right) = \int_0^t \mathbf{E}(W_t W_u) \, du = \int_0^t u \, du = \frac{t^2}{2}, \quad (11.24)$$

$$\mathbf{E}\left(W_t \int_0^t u \, dW_u\right) = \mathbf{E}\left(W_t \left(tW_t - \int_0^t W_u \, du\right)\right) = t^2 - \frac{t^2}{2} = \frac{t^2}{2}, \quad (11.25)$$

$$\mathbf{E}\left(W_t \int_0^t W_u^2 \, dW_u\right) = \mathbf{E}\left(W_t \left(\frac{W_t^3}{3} - \int_0^t W_u \, du\right)\right) = \frac{t^2}{2}. \quad (11.26)$$

In the final derivation we have exploited the Gaussian factorization property of the Brownian process $\langle W_t^4 \rangle = 3 \langle W_t^2 \rangle^2$, and thus re-expressed the integral over the Wiener measure.[28] Altogether we find that

$$\mathbf{E}\left((\Delta \hat{x}_t)^2\right) = \sigma^2 t + \left(\left(\sigma^{(1)}\sigma\right)^2 + 2F^{(1)}\sigma^2 + 2F\sigma\sigma^{(1)} + \sigma^{(2)}\sigma^3\right)\frac{t^2}{2}. \quad (11.27)$$

We treat the corresponding analysis for a vector process in Appendix G.

11.2 Rayleigh and gamma processes

Having generated an iterative solution to a generic SDE with multiplicative noise, we now apply these results to SDEs describing processes of particular interest in the study of non-Gaussian scattering processes.

First, we recall the Rayleigh process, described by the scalar SDE

$$\mathrm{d}z_t = (1 - z)\,\mathrm{d}t + \sqrt{2z_t}\mathrm{d}W_t. \quad (11.28)$$

The relationship between this SDE and other descriptions of the intensity of a Gaussian speckle process is discussed in Tough (1987). To make contact with the general results above, we make the identifications

$$\begin{cases} F(z_0) = (1 - z_0), \\ F^{(1)}(z_0) = -1, \\ F^{(2)}(z_0) = 0 \end{cases} \quad (11.29)$$

$$\begin{cases} \sigma(z_0) = \sqrt{2z_0}, \\ \sigma^{(1)}(z_0) = \frac{1}{2}\sqrt{\frac{2}{z_0}}, \\ \sigma^{(2)}(z_0) = -\frac{1}{4}\sqrt{\frac{2}{z_0^3}}. \end{cases}$$

When these are substituted into (11.19) we find that

$$\begin{aligned} \mathbf{E}(\Delta z) &= Ft + \frac{t^2}{2}FF^{(1)} + \frac{1}{4}F^{(2)}\sigma t^2 + O\left(t^3\right) \\ &= (1 - z_0)t\left(1 - \frac{t}{2}\right) + O\left(t^3\right). \end{aligned} \quad (11.30)$$

The mean square value of the stochastic part of the increment in z can be evaluated as follows

---

[28] This can be derived by calculating the stochastic differential $\mathrm{d}(W_t^4)$, integrating, and taking the expected value.

$$\mathbf{E}\left((\Delta \hat{z})^2\right) = \sigma^2 t + \left(\left(\sigma^{(1)}\sigma\right)^2 + 2F^{(1)}\sigma^2 + 2F\sigma\sigma^{(1)} + \sigma^{(2)}\sigma^3\right)\frac{t^2}{2} + O\left(t^3\right)$$
$$= 2z_0 t + (1 - 3z_0)t^2 + O\left(t^3\right). \tag{11.31}$$

In the special case of the Rayleigh process described by (11.28) further progress can be made, albeit by eschewing the iterative approach. Instead, we identify the FPE that is stochastically equivalent to this SDE and solve explicitly for its propagator. Using this we can evaluate expectations of the form

$$\mathbf{E}\left[(z_t - z_o)\right], \quad \mathbf{E}\left[(z_t - z_o)^2\right] \tag{11.32}$$

for arbitrary time intervals t. By making a suitable small time expansion we can check the validity of our more general analysis in this case. The statistics of this increment are of interest in the context of improving the precision in the stochastic volatility analyses of Chapter 12; the results below that hold for arbitrary times identify precision corrections to the $O(t)$ treatment that we have in practice applied to the experimental data analysis, which serves to illustrate the theory sufficiently well. However, in situations that present themselves where the sample time is not as small as one would like, the higher order correction analysis we provide here should be useful.

We recall that the FPE equivalent to (11.28) is

$$\frac{\partial^2}{\partial z^2}\left(zP(z,t|z_0)\right) + \frac{\partial}{\partial z}\left((z-1)P(z,t|z_0)\right) = \frac{\partial}{\partial t}P(z,t|z_0) \tag{11.33}$$

whose fundamental solution or propagator satisfies the initial condition

$$P(z,0|z_0) = \delta(z - z_0). \tag{11.34}$$

(Further background information on this topic can be found in Tough 1987; our notation follows that of this reference quite closely.) The propagator can be expanded in terms of the Laguerre polynomials as

$$P(z,t|z_0) = \exp(-z)\sum_{n=0}^{\infty} L_n(z) L_n(z_0)\exp(-nt) \tag{11.35}$$

$$L_n(z) = \frac{1}{n!}\exp(z)\left(\frac{\mathrm{d}}{\mathrm{d}z}\right)^n \left(\exp(-z)z^n\right).$$

Using this expansion, and some basic properties of these polynomials, we find that

$$\mathbf{E}(z_t - z_0) = \int_0^{\infty}(z - z_0)P(z,t|z_0)\,\mathrm{d}z$$
$$= (1 - z_0)(1 - \exp(-t))$$
$$= (1 - z_0)\left(t - \frac{t^2}{2} + \frac{t^3}{6}\right) + O\left(t^4\right) \tag{11.36}$$

$$\mathbf{E}\left((z_t - z_0)^2\right) = \int_0^\infty (z - z_0)^2 P(z,t|z_0)\, dz$$

$$= z_0^2 \left(1 - \exp(-t)\right)^2 + z_0 \left(4\exp(-t) - 2\right)\left(1 - \exp(-t)\right) + 2\left(1 - \exp(-t)\right)^2$$

$$= 2z_0 t + \left(2 - 5z_0 + z_0^2\right) t^2 - \left(2 - \frac{13}{3} z_0 + z_0^2\right) t^3 + O\left(t^4\right). \tag{11.37}$$

Higher order moments can, in principle, be determined from the characteristic function, which is most conveniently derived using an alternative form for the propagator

$$P(z,t|z_0) = \frac{1}{1-\theta} \exp\left(-\frac{(z + z_0 \theta)}{1-\theta}\right) I_0 \left(\frac{2\sqrt{\theta z z_0}}{1-\theta}\right),$$

$$\theta = \exp(-t). \tag{11.38}$$

Here I_0 is a zeroth-order modified Bessel function of the first kind. Using this we find that

$$\mathbf{E}(\exp(-s(z_t - z_0))) = \sum_{n=0}^\infty \frac{(-s)^n \mathbf{E}\left((z_t - z_0)^n\right)}{n!}$$

$$= \frac{1}{1-\theta} \exp\left(z_0 \left(s + \frac{\theta}{1-\theta}\right)\right) \int_0^\infty \exp\left(-z \left(s + \frac{1}{1-\theta}\right)\right) I_0 \left(\frac{2\sqrt{\theta z z_0}}{1-\theta}\right) dz$$

$$= \frac{\exp\left(\frac{s z_0 (1+s)(1-\theta)}{1 + s(1-\theta)}\right)}{1 + s(1-\theta)}. \tag{11.39}$$

Expanding this characteristic function in a power series and equating coefficients of powers of s verifies the results (11.36) and (11.37). These in turn allow us to verify the results obtained to order t^2 by iteration. In particular, the mean square of the stochastic part of the increment can be calculated as

$$\mathbf{E}\left(\Delta \hat{z}^2\right) = \mathbf{E}\left((z_t - z_0)^2\right) - (1 - z_0)^2 t^2 + O\left(t^3\right)$$

$$= 2 z_0 t + (1 - 3 z_0) t^2 + O\left(t^3\right). \tag{11.40}$$

The results we have presented in this section demonstrate the application of the general formulae derived earlier, and provide us with some further insights into the SDE description of the K-scattering process. In the special case of the Rayleigh process, we have been able to calculate equivalent quantities in more detail by making recourse to the closed form solution of the corresponding FPE that is available in this case. Limiting forms of these results provide verification of our more general iterative analysis. In Section 11.3, we discuss briefly how the results we have derived for the K-scattering process can be exploited in a numerical simulation algorithm.

So far in this chapter we have examined the iterative solutions of some SDEs that occur in the description of clutter processes. We have considered the case where the SDEs incorporates multiplicative noise, and so has been interpreted within the Ito convention. A corresponding simpler treatment holds for the case of iterative solutions of ordinary differential equations and Langevin equations incorporating additive noise. The greater complexity of the multiplicative noise calculations has limited us to results valid up to second order only, whereas the corresponding Langevin case would naturally yields results valid up to third order in time. This fact highlights some of the problems encountered in the analysis of systems with multiplicative, as opposed to additive, noise.

The general results obtained have been specialized to the relatively simple SDEs that describe the Rayleigh and K-scattering processes. As these incorporate multiplicative noise, we have generated results valid up to second order in time. Recent experimental work (Field and Tough 2003a) has focussed attention of the behaviour of increments in the intensity of the Rayleigh and K processes that occur over time intervals that are very short in comparison to the correlation timescale of the process concerned. Accordingly, this analysis can be carried out using a first-order expansion in time. The results we present here are useful in assessing the validity of this approximation in cases where the relevant sample time interval, for example as dictated by experimental design, is a significant proportion of the correlation time. In the special case of the Rayleigh process we are able to make significantly more progress, by adopting the Fokker–Planck formulation of the problem. The results we derive specialize to provide a verification of the analysis for more general processes, and should be of relevance to analyses of empirical data with shorter correlation timescales and constraints on the available sample frequency ranges.

In addition to providing some insight into the temporal development of the system described by SDEs incorporating multiplicative noise, the results of this iterative analysis also allow us to simulate their behaviour numerically, in effect providing numerical solution to the SDEs concerned. One way to achieve this, which has been applied to Langevin type equations (Greenside and Helfand 1981) is very much in the spirit of the Runge–Kutta algorithm, frequently applied to the solution of ordinary differential equations. These workers develop a method of stepping forward in time, combining several contiguous values of the dependent variables in such a way that agreement with a Taylor series development of the solution is maintained to some stated order in the time step (Whittaker and Watson 1969). When applying this method to SDEs, both the deterministic and stochastic parts of the solution have to be controlled. Given the difficulties encountered in characterizing the stochastic part of the solution, it is reasonable to require agreement to second order in the time increment; typically Runge–Kutta algorithms applied to deterministic equations seek agreement to fourth or even higher order. We shall apply these methods to the solution of the SDEs discussed in Section 11.3, and so provide a more accurate method for the simulation of K-scattering processes.

11.3 Compound K-model

In Sections 11.1 and 11.2, we described how stochastic processes described by SDEs with multiplicative noise can be simulated by matching the mean and mean square increments generated in a Runge–Kutta-like algorithm with those obtained from a systematic iterative solution of the SDE. The computational complexity restricts the accuracy of these algorithms to $O\big((\Delta t)^2\big)$. Thus, in a situation where two coupled processes evolve with widely differing de-correlation times (e.g. for $\mathcal{A} \ll \mathcal{B}$ in Theorem 8.1) the algorithm will be accurate only to $O\big((\Delta t)^2\big)$, even though $\mathcal{A}^3 (\Delta t)^3$ and higher order terms may be negligibly small. In the case of the K-scattering model of Chapter 8, in a radar context the so-called 'compound K model', it is possible to overcome this limitation.

As the compound form represents the K-scattering process as a product of two independent processes, described as in Section 8.1, one might suppose that the characteristic time of the Rayleigh process (i.e. the reciprocal of the constant $\mathcal{B}$) would define the appropriate integration time step. However, it proves to be possible to propagate the Rayleigh process along over arbitrary time steps, using the following artifice. We have seen that the Rayleigh process of unit power is obtained by summing the squares of two component Ornstein–Uhlenbeck processes. Each of these is described by a *linear* SDE and so can be integrated explicitly over any time step. Thus it is only the slower gamma modulation process that limits the step size, as the overall algorithm has an accuracy of $O\big((\mathcal{A}\Delta t)^2\big)$.

The coupled SDEs we solve to simulate the complex Ornstein–Uhlenbeck process are

$$\mathrm{d}p = -\tfrac{1}{2}p\mathrm{d}t - \omega q \mathrm{d}t + \tfrac{1}{\sqrt{2}}\mathrm{d}W_p(t),$$
$$\mathrm{d}q = -\tfrac{1}{2}q\mathrm{d}t + \omega p \mathrm{d}t + \tfrac{1}{\sqrt{2}}\mathrm{d}W_q(t). \qquad (11.41)$$

We can write these as a linear matrix equation

$$\mathrm{d}(\exp(\mathbf{A}t) \cdot \mathbf{u}) = \tfrac{1}{\sqrt{2}} \exp(\mathbf{A}t)\,\mathrm{d}\mathbf{B}(t),$$
$$\mathbf{u} = \begin{pmatrix} p \\ q \end{pmatrix},$$
$$\mathbf{A} = \begin{pmatrix} 1/2 & \omega \\ -\omega & 1/2 \end{pmatrix} = \tfrac{1}{2}\mathbf{1} + \omega \mathbf{P}, \qquad (11.42)$$
$$\mathbf{B}(t) = \begin{pmatrix} W_p(t) \\ W_q(t) \end{pmatrix}.$$

This we can integrate to yield

$$\mathbf{u}(t) = \exp(-\mathbf{A}t) \cdot \mathbf{u}(0) + \int_0^t \exp(-\mathbf{A}(t-t')) \cdot \mathrm{d}\mathbf{B}(t'). \qquad (11.43)$$

The mean of this process is given by

$$\langle \mathbf{u}(t) \rangle = \exp(-\mathbf{A}t) \cdot \mathbf{u}(0). \tag{11.44}$$

The exponential matrix can be evaluated, exploiting the trivial commutation of **1** and **P** and **P**'s identification as the matrix square root of the identity matrix, $\mathbf{P} \cdot \mathbf{P} = -\mathbf{1}$. Thus we find that

$$\begin{aligned}\exp(\mathbf{A}t) &= \exp(\tfrac{t}{2}) \begin{pmatrix} \cos\omega t & \sin\omega t \\ -\sin\omega t & \cos\omega t \end{pmatrix}, \\ (\exp(\mathbf{A}t))^{\mathrm{T}} &= \exp(\tfrac{t}{2}) \begin{pmatrix} \cos\omega t & -\sin\omega t \\ \sin\omega t & \cos\omega t \end{pmatrix}.\end{aligned} \tag{11.45}$$

We can now evaluate the covariance matrix of the process as follows:

$$\begin{aligned}&\left\langle (\mathbf{u}(t) - \langle \mathbf{u}(t)\rangle)^{\mathrm{T}}(\mathbf{u}(t) - \langle \mathbf{u}(t)\rangle) \right\rangle \\ &= \frac{1}{2} \left\langle \int_0^t \int_0^t (\exp(-\mathbf{A}(t-t_1)) \cdot \mathrm{d}\mathbf{B}(t_1)))^{\mathrm{T}} (\exp(-\mathbf{A}(t-t_2)) \cdot \mathrm{d}\mathbf{B}(t_2))) \right\rangle \\ &= \frac{1}{2} \int_0^t \exp(-\mathbf{A}(t-t_1)) \cdot (\exp(-\mathbf{A}(t-t_1)))^{\mathrm{T}} \mathrm{d}t_1 \\ &= \frac{1}{2} \mathbf{1} \int_0^t \exp(-t_1) \, \mathrm{d}t_1 = \frac{1}{2}(1 - \exp(-t))\,\mathbf{1}.\end{aligned} \tag{11.46}$$

Here we have exploited the Ito isometry (see Section 2.1.1) to evaluate the expectation. Using these results, we can step the complex Ornstein–Uhlenbeck (COU) process through *arbitrary* time intervals via

$$\begin{pmatrix} p(t_n) \\ q(t_n) \end{pmatrix} = \exp\left(\frac{-(t_n - t_{n-1})}{2}\right) \begin{pmatrix} \cos(\omega(t_n - t_{n-1})) & -\sin(\omega(t_n - t_{n-1})) \\ \sin(\omega(t_n - t_{n-1})) & \cos(\omega(t_n - t_{n-1})) \end{pmatrix}$$
$$\times \begin{pmatrix} p(t_{n-1}) \\ q(t_{n-1}) \end{pmatrix} + \sqrt{\frac{(1 - \exp(-(t_n - t_{n-1})))}{2}} \begin{pmatrix} g_p \\ g_q \end{pmatrix}. \tag{11.47}$$

in which g_p, g_q are independent Gaussian random variables drawn from a distribution with zero mean and a unit variance. This representation of the COU process can be used to generate a unit mean Rayleigh intensity process through

$$u(t_n) = p(t_n)^2 + q(t_n)^2. \tag{11.48}$$

The Doppler frequency can be suppressed if we wish only to generate the intensity; in these circumstances we have

$$\begin{aligned}p(t_n) &= \exp\left(\tfrac{-(t_n - t_{n-1})}{2}\right) p(t_{n-1}) + \sqrt{\tfrac{(1 - \exp(-(t_n - t_{n-1})))}{2g_p}}, \\ q(t_n) &= \exp\left(\tfrac{-(t_n - t_{n-1})}{2}\right) q(t_{n-1}) + \sqrt{\tfrac{(1 - \exp(-(t_n - t_{n-1})))}{2g_q}}.\end{aligned} \tag{11.49}$$

This unit power Rayleigh process can be simulated over arbitrary time steps. To yield a K-scattering process it should be multiplied by the correlated gamma process, generated by the solution of (8.9) which will typically have accuracy to $O\left((\mathcal{A}\Delta t)^2\right)$ characteristic of the entire compound process. Observe that the procedure we have outlined here provides us with a relatively efficient method for the simulation of the complex amplitude of the scattered field, arising as the product according to (8.3).

In this section we have seen that the compound representation of the K-scattering process also facilitates its (computer) simulation. In particular, it is possible to construct an algorithm whose accuracy depends only on the ratio of the time step of numerical integration to the long time characteristic of the decay of the correlation with respect to the constituent gamma process.

11.4 Influence of Doppler on volatilities

The anomaly detection method described in Chapter 12 is based on the determination of the volatility of the complex field, based on the measurement change in the returned complex field over short periods of time. In the radar data analysed thus far this time step is determined by the pulse repetition frequency, and is of the order of 1 ms. The returned field develops through diffusive and Doppler effects; it is the former that is characterized by the volatility. For the determination of the volatility to be accurate, Doppler effects should be negligible over the timescale of the measurement. We will now consider whether this is the case.

To compare the effects of Doppler and diffusion we consider the simple model of a COU process, as discussed in Section 8.1. The underlying SDE is

$$\begin{aligned} \mathrm{d}p &= -\frac{1}{2}\beta p \mathrm{d}t - \omega q \mathrm{d}t + \sqrt{\frac{\beta}{2}} \mathrm{d}W_t^{(p)}, \\ \mathrm{d}q &= -\frac{1}{2}\beta q \mathrm{d}t + \omega p \mathrm{d}t + \sqrt{\frac{\beta}{2}} \mathrm{d}W_t^{(q)}. \end{aligned} \quad (11.50)$$

The SDE associated with the intensity $z = p^2 + q^2$ is obtained using the usual rules of Ito calculus as

$$\mathrm{d}z = \beta(1-z)\,\mathrm{d}t + \sqrt{2\beta z}\mathrm{d}W_t^{(z)}. \quad (11.51)$$

This is stochastically equivalent to the FPE

$$\frac{\partial P(z,t)}{\partial t} = \beta \frac{\partial}{\partial z}\left((z-1)P(z,t)\right) + \beta \frac{\partial^2}{\partial z^2}\left(zP(z,t)\right) \quad (11.52)$$

which has the stationary solution $P(z) = \exp(-z)$. The correlation function of the intensity implicit in this model exhibits a simple exponential decay, falling off as $\exp(-\beta t)$. Because the SDEs for the COU process are linear they can be integrated over an arbitrary time step, as explained earlier. Thus we have

$$p_0 = p(0), \quad q_0 = q(0);$$
$$\Delta p = p(t) - p(0); \quad \Delta q = q(t) - q(0)$$
$$\begin{pmatrix} \Delta p \\ \Delta q \end{pmatrix} = \left(\exp(-\beta t/2) \begin{pmatrix} \cos\omega t & -\sin\omega t \\ \sin\omega t & \cos\omega t \end{pmatrix} - \begin{pmatrix} 1 & 0 \\ 0 & 1 \end{pmatrix} \right) \begin{pmatrix} p_0 \\ q_0 \end{pmatrix}$$
$$+ \sqrt{\frac{(1 - \exp(-\beta t))}{2}} \begin{pmatrix} g_p \\ g_q \end{pmatrix}. \tag{11.53}$$

The g-functions are drawn independently from a zero mean, unit variance Gaussian distribution. These results form the basis of a simple simulation of coherent clutter, which is discussed in more detail in the next section.

To establish the behaviour of the square modulus of the increment in the complex field $\Psi = p + iq$, we make an expansion of $\langle (\Delta p)^2 + (\Delta q)^2 \rangle$; here the angular bracket denotes an averaging over the gs. In this way, we find that

$$\langle (\Delta p)^2 + (\Delta q)^2 \rangle = 1 - \exp(-\beta t) + (p_0^2 + q_0^2)(1 + \exp(-\beta t)$$
$$- 2\exp(-\beta t/2)\cos\omega t). \tag{11.54}$$

We can expand this up to second order in t as

$$\langle (\Delta p)^2 + (\Delta q)^2 \rangle \sim \beta t + \beta^2 t^2 \left(-\frac{1}{2} + \frac{z_0}{4} + \frac{\omega^2 z_0}{\beta^2} \right), \tag{11.55}$$
$$z_0 = p_0^2 + q_0^2.$$

To ascertain the effects of the Doppler evolution on the measurement of the volatility in the complex field (which we can see from (11.55) to be equal to β), we must introduce estimates of the various terms in this expression.

An independent analysis of the data presented in Section 12.2 (Ward 2002), for which the time step t is known to be 1 ms, indicates that the intensity correlation function decays in the order of 10 ms; thus we approximate $\beta \approx 100$. The Doppler frequency is identified as 100 Hz, i.e. an angular frequency $\omega = 200\pi$. When these values are introduced into (11.55) we find that

$$\frac{\langle |\Delta\Psi|^2 \rangle}{\beta t} = 1 + 0.1 \left(-\frac{1}{2} + \left(4\pi^2 + \frac{1}{4} \right) z_0 \right)$$
$$\approx 1 + 4z_0. \tag{11.56}$$

From this we see that, for the data analysed in Field and Tough (2003a) and Section 12.2, Doppler induced effects are comparable with the diffusive evolution of the complex field and so can influence the determination of the volatility. It is particularly striking that the intensity dependence of the Doppler term is the same as that interpreted as a contribution to the volatility of the type previously accounted for by the endogenous model. The interaction of the diffusive, Doppler, and power modulation terms and their effects on the performance of the anomaly detection can best be assessed by simulation studies. Methods with which these may be implemented will be discussed in the following.

11.5 Coherent clutter

In the previous section, we analysed the COU process to gain some insight into the procedure adopted in the measurement of volatilities. We will now show how the COU process provides us with a useful simulation tool that allows us to reproduce many of the features observed in the data analysis underpinning the volatility-based anomaly detector. When combined multiplicatively with the square root of a gamma process, whose simulation is described in the following section, the COU and related processes should provide us with a valuable simulation tool for the investigation of the performance of the volatility based anomaly detector described in Section 12.2.

The simulation procedure implicit in (11.53) is trivial to implement and represents the extension to the complex domain of a method described, for example, by Pike and co-workers (Hughes *et al.* 1973) some thirty or more years ago.[29] The effect of a Doppler term on the evolution of the returned field, represented as a complex number in the Argand plane, can be illustrated as follows. Figures 11.1–11.3 show this evolution for zero, small and significant Doppler respectively; in the last of these the relative contributions of the diffusive and Doppler effects are much the same as in the data analysis discussed in the previous section. The qualitative differences between these plots are quite evident.

We now assess the extent to which the complex field volatility, measured over a single time step, tracks the intensity of the field. In Fig. 11.4, we display traces of the intensity (solid) and measured field volatility (dotted) derived from simulated COU data for which $\beta = 1.0$, $\omega = 6.0$, $\Delta t = 0.1$. The plots have been scaled so that the range of each is the same; a strong correlation between the intensity and measured field volatility is evident in this plot. In Fig. 11.5, however, we see that this correlation is greatly reduced when the COU process has no Doppler component. This simple simulation, for typical radar parameter values, highlights the care that is needed in the interpretation of measured field amplitude volatilities, in that the terms of $O(\delta t^2)$ may have a significant contribution. Thus the results of Section 12.2 depend on the presence of Doppler for such high degrees of correlation in the clutter domain. Measurements of the intensity volatility can be analysed similarly. Figure 11.6 shows the correlation between the intensity (solid) and measured intensity volatility (dotted) derived from simulated COU data for which there is no Doppler component. Once again we have scaled the plots to facilitate their comparison. Thus, unlike the case of amplitude volatilities above, the situation as regards correlation for the intensity volatility is unaffected by Doppler, as anticipated since the Doppler transformation, consisting of rigid rotations of the Argand plane representing the amplitude, leaves the intensity process invariant.[30]

---

[29] A *Mathematica* notebook that implements this and other simulation procedures is available from the author.

[30] This fact can be readily established from computing the intensity SDE from the amplitude SDE containing the Doppler term as provided by (8.60).

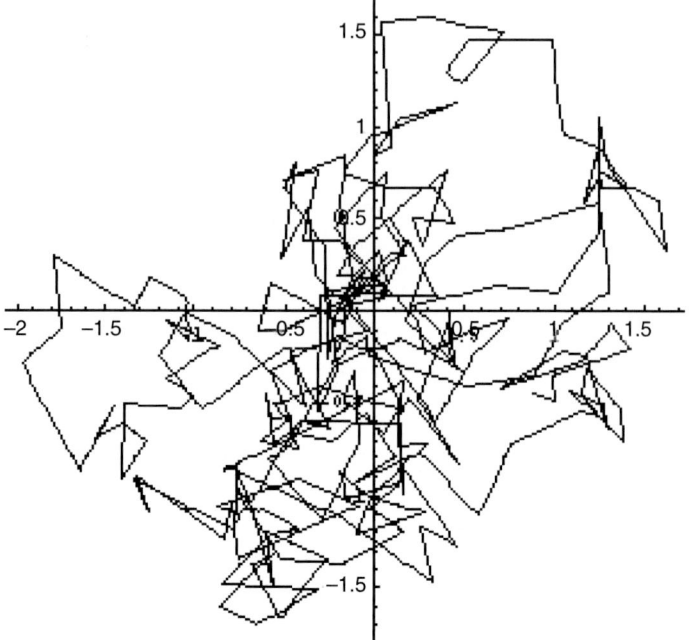

FIG. 11.1. Realization of a COU process; $\beta = 1.0$, $\omega = 0$, $\Delta t = 0.1$.

11.5.1 *Target returns in clutter*

A simple model of a 'target plus clutter' (radar) return can be constructed by adding a coherent signal of constant amplitude and prescribed Doppler frequency directly to the COU process. It is interesting to note that a simple FFT displays even a fairly weak target signal quite prominently, as long as there is a significant difference between the Doppler frequencies characteristic of the clutter and target. The introduction of fluctuations into the target return amplitude, perhaps modelled by a gamma or similar process, might yield a more realistic model. Figure 11.7 shows an (I, Q) plot of this simple target plus clutter model; the Fourier transform of this data is shown in Fig. 11.8, and demonstrates the efficacy of simple Doppler processing in this case.

In this section we have assembled the components of a simulation capability that should allow us to assess the performance of the volatility-based anomaly detector described in Section 12.2 [*orig.* Field and Tough (2003a)]. An analysis of the impact of the Doppler characteristics of the returned signal on the measurement of the field volatility indicates that some care will be necessary in the separation and identification of the processes involved. This also highlights the need for further controlled simulation and real-data-based studies of the anomaly detector, to establish its robustness and general applicability.

Some simple preliminary simulations based on the COU process confirm our findings on the influence of Doppler on measured volatilities. The incorporation

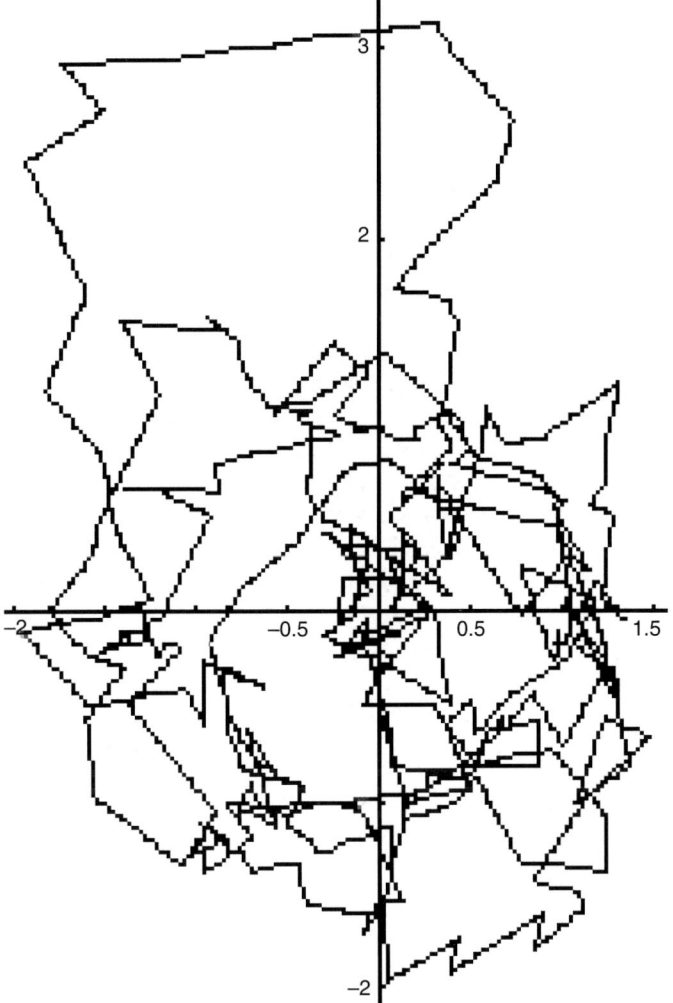

FIG. 11.2. Realization of a COU process; $\beta = 1.0$, $\omega = 1.0$, $\Delta t = 0.1$.

of a coherent target return into this model is straightforward, and allows us to demonstrate the effectiveness of simple FFT processing in isolating the target when its Doppler characteristics differ significantly from those of the clutter.

11.6 Second-order algorithms

The endogenous model of the clutter process represents the returned field as the product of a COU process with a stochastically varying amplitude, i.e. the square root of a gamma process. Thus, to complete our simulation capability,

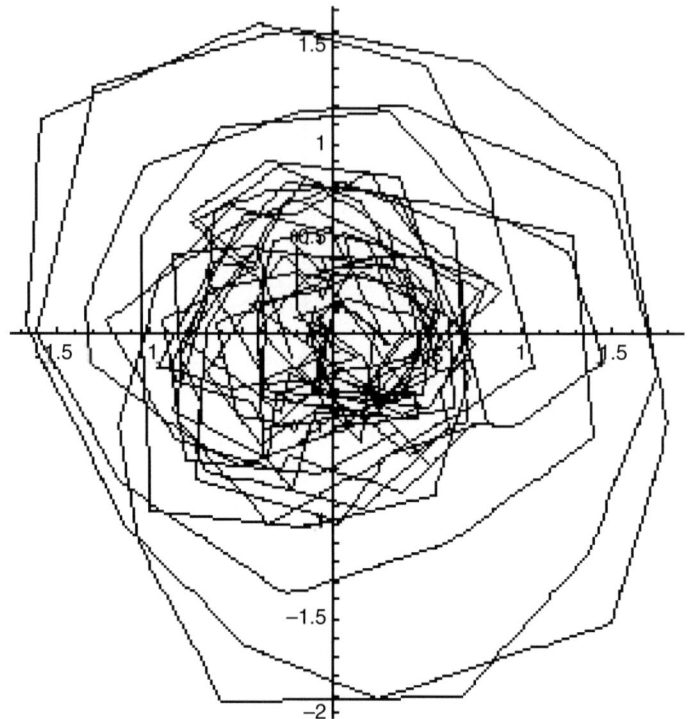

FIG. 11.3. Realization of a COU process; $\beta = 1.0$, $\omega = 6.0$, $\Delta t = 0.1$.

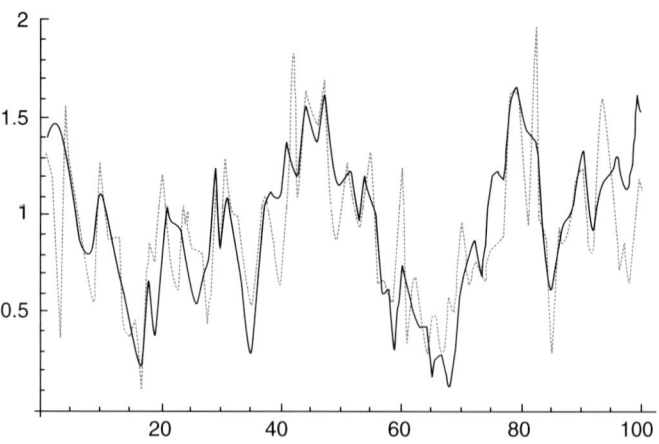

FIG. 11.4. Comparison of the intensity (solid) and measured field volatility (dotted) of a COU process with $\beta = 1.0, \omega = 6.0, \Delta t = 0.1$.

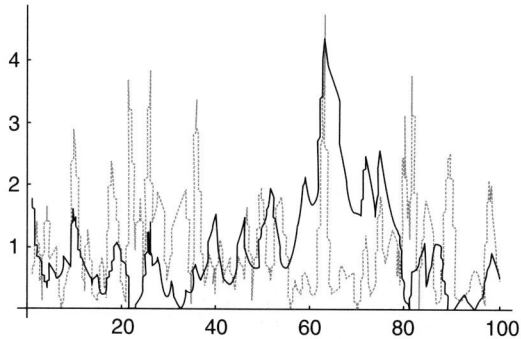

FIG. 11.5. Comparison of the intensity (solid) and measured field volatility (dotted) of a COU process with $\beta = 1.0, \omega = 0, \Delta t = 0.1$.

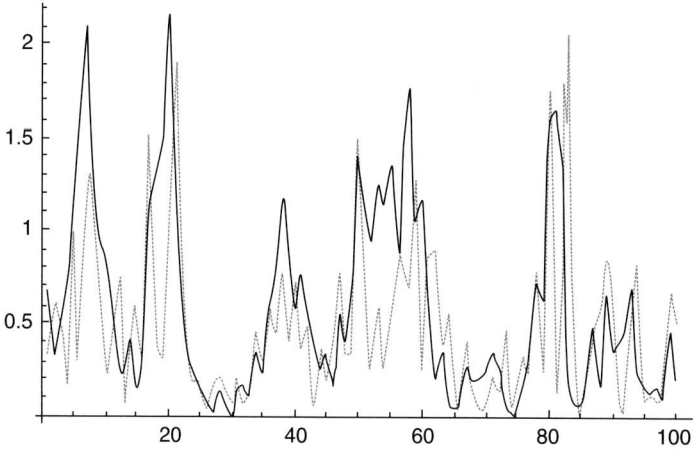

FIG. 11.6. Comparison of the intensity (solid) and measured intensity volatility (dotted) of a COU process with $\beta = 1.0, \omega = 0, \Delta t = 0.1$.

we must devise an effective method for the simulation of such a process. As we have discussed previously in Section 8.1, the temporal evolution of a gamma process x_t is described by the SDE (8.9). We observe that the volatility term depends on the current value of x_t, i.e., it is an instantaneous *stochastic* volatility.[31] Closed form solutions of SDEs of this type are not generally available; the initial temporal evolution of the process can, nonetheless, be characterized by a systematic iteration of the SDE, re-expressed in an integral form. A heuristic development of this solution was given by Pusey and Tough (1982), in a discussion presented in the context of the Brownian dynamics of hydrodynamically interacting particles. A more rigorous and detailed analysis of the problem was

---

[31] The reader should compare the discussion of models in mathematical finance in Appendix D.

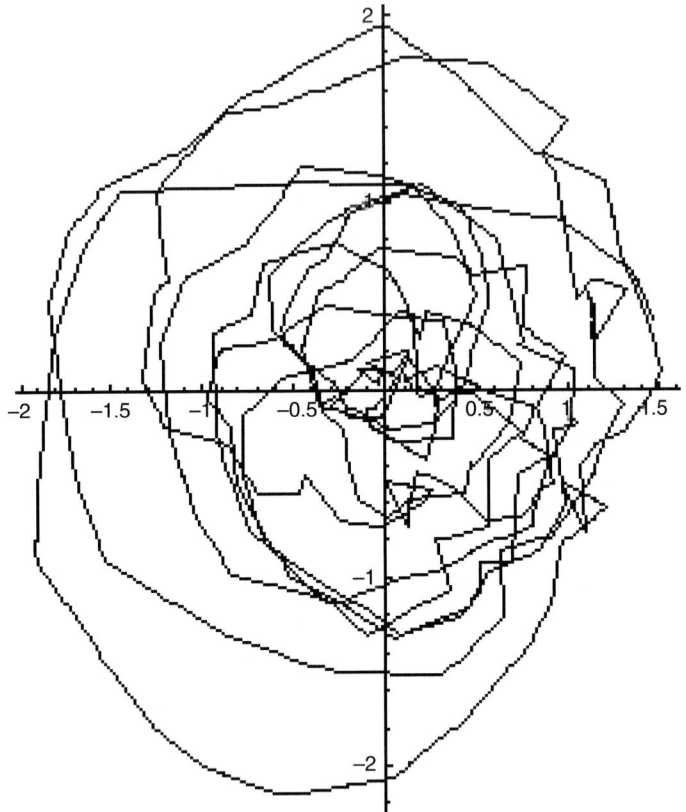

FIG. 11.7. An *IQ* plot of a target plus clutter return; the target power is one quarter that of the clutter, while the target Doppler is twice that of the clutter.

given by Klauder and Petersen (1985), whose results have been re-derived recently by Quiang and Habib (2000). In essence, these workers determined the mean and variance of the increment in a process described by a SDE, to second order in the time step and expressed in terms of the drift and volatility functions and their derivatives.

A simple simulation of the gamma process proceeds as follows. Integrating (8.9) we obtain the simple recursion

$$x(n+1) = x(n) + \mathcal{A}(\nu - x(n))\Delta t + \sqrt{2\mathcal{A}x(n)\Delta t}\,g(n). \quad (11.57)$$

Here Δt is the time step over which the process evolves and the $g(n)$ are independent Gaussian distributed random variables, with zero mean and unit variance. This simple procedure reproduces the mean and variance of the increment in the process to linear order in Δt, and is frequently referred to as an Euler algorithm.

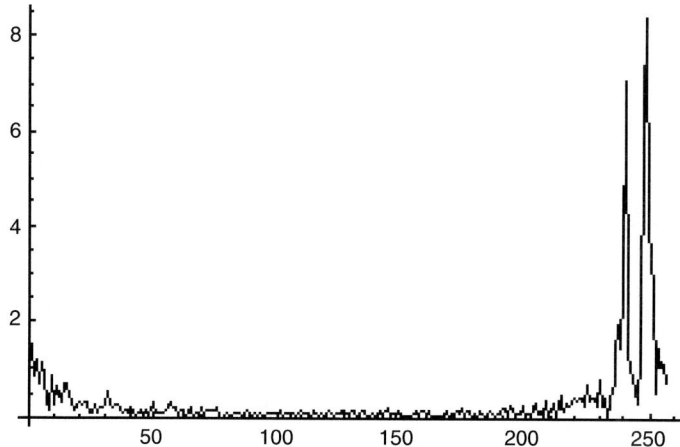

FIG. 11.8. Amplitude of Fourier transform of the preceding data, demonstrating the isolation of the coherent target from the clutter.

(Much of the greatest part of the solution of SDEs by numerical simulation is based on this method, as in Quiang and Habib.) Increments in x whose mean and variance are correct to quadratic order in Δt could be constructed using the algorithms presented in the latter two references. Klauder and Petersen display considerable ingenuity to obviate the need for evaluation of derivatives of the drift and volatility functions, basing their algorithm on the classic Runge–Kutta algorithm for the solution of ordinary differential equations (e.g. Fox and Mayers 1968). However, in the case of the gamma process these drift and volatility functions are particularly simple; thus it is possible to characterize the increments in the process directly, rather than by iteration. In this special case, we can therefore produce an algorithm, accurate to second order in Δt, that is almost as simple to implement as the Euler algorithm (11.57).

Recall from Chapter 3 that the FPE stochastically equivalent to (8.9) is

$$\frac{\partial}{\mathcal{A}\partial t}P(x,t|x_0) = \frac{\partial}{\partial x}\left((x-\nu)P(x,t|x_0)\right) + \frac{\partial^2}{\partial x^2}\left(xP(x,t|x_0)\right). \quad (11.58)$$

The propagator of this can be constructed in a reasonable closed form[32] as

$$P(x,t|x_0) = \frac{1}{1-\exp(-\mathcal{A}t)}\left(\frac{x\exp(\mathcal{A}t)}{x_0}\right)^{(\nu-1)/2}\exp\left(-\frac{(x+x_0\exp(-\mathcal{A}t))}{1-\exp(-\mathcal{A}t)}\right)$$
$$\times I_{\nu-1}\left(\frac{2\exp(-\mathcal{A}t/2)\sqrt{xx_0}}{1-\exp(-\mathcal{A}t)}\right). \quad (11.59)$$

In the absence of the stochastic driving term, (8.9) reduces to

---

[32]Here $I_{\nu-1}$ is a modified Bessel function of the first kind (see Section 9.6 in Abramowitz and Stegun 1970).

which can be solved to yield

$$\begin{aligned} x_D &= \nu - (\nu - x_0)\eta, \\ x_D &\equiv x_D(t), \quad x_0 \equiv x(0), \quad \eta = \exp(-\mathcal{A}t). \end{aligned} \tag{11.61}$$

Here, the subscript D denotes the deterministic part of x. We can characterize the stochastic part of the development of x by evaluating the generating function

$$C(s) = \int_0^\infty \exp(-s(x - x_D)) P(x, t|x_0)\, dx. \tag{11.62}$$

A development of $C(s)$ in powers of s then allows us to identify the required expectation values

$$C(s) = \sum_{n=0}^\infty \frac{(-1)^n}{n!} s^n \langle (x - x_D)^n \rangle. \tag{11.63}$$

Recalling that $\eta = \exp(-\mathcal{A}t)$, and on substituting (11.59) into (11.62), we find that

$$C(s) = \frac{\exp\left((1-\eta)s(\nu + \eta s x_0/(1+(1-\eta)s))\right)}{(1+(1-\eta)s)^\nu}. \tag{11.64}$$

Equation (11.64) can now be expanded as a power series in s to give

$$C(s) = 1 + \tfrac{1}{2}(1-\eta)((1-\eta)\nu + 2\eta x_0)s^2 - \tfrac{1}{3}(1-\eta)^2((1-\eta)\nu + 3\eta x_0)s^3 + \cdots. \tag{11.65}$$

We note that the term linear in s vanishes; the mean value of the stochastic part of x vanishes identically, as expected. We can now read off the following:

$$\begin{aligned} \langle (x - x_D) \rangle &= 0, \\ \left\langle (x - x_D)^2 \right\rangle &= (1-\eta)[(1-\eta)\nu + 2\eta x_0], \\ \left\langle (x - x_D)^3 \right\rangle &= 2(1-\eta)^2[(1-\eta)\nu + 3\eta x_0]. \end{aligned} \tag{11.66}$$

Finally, we introduce the above expression for η into these results, and expand up to cubic order in the time step, which leads us to

$$\begin{aligned} \langle (x - x_D) \rangle &= 0, \\ \left\langle (x - x_D)^2 \right\rangle &= 2x_0\mathcal{A}t + (\nu - 3x_0)(\mathcal{A}t)^2 + \left(\frac{7x_0}{3} - \nu\right)(\mathcal{A}t)^3 + \cdots, \\ \left\langle (x - x_D)^3 \right\rangle &= 6x_0(\mathcal{A}t)^2 + 2(6x_0 - \nu)(\mathcal{A}t)^3 + \cdots. \end{aligned} \tag{11.67}$$

It is interesting to note that the mean cube of the stochastic part of x has a *quadratic* time contribution; in earlier work (Klauder and Petersen 1985; Quiang and Habib 2000; Pusey and Tough 1982) this term has not been analysed explicitly for SDEs with multiplicative noise, and has been assumed to be third order in time.

If we now expand x_D to second order in time as

$$x_D = x_0 + (\nu - x_0)\mathcal{A}t(1 - \mathcal{A}t/2) \tag{11.68}$$

we see that we can construct an algorithm similar in structure to (11.57) that nonetheless generates increments whose mean and variance are maintained to quadratic order in the time step:

$$\begin{aligned} x(n+1) = x(n) + \mathcal{A}(\nu - x(n))\Delta t \left(1 - \frac{\mathcal{A}\Delta t}{2}\right) \\ + \sqrt{2\mathcal{A}x(n)\Delta t + (\nu - 3x(n))(\mathcal{A}\Delta t)^2}\, g(n). \end{aligned} \tag{11.69}$$

One of the problems faced when generating a gamma Markov process by simulation of the solution of the SDE (8.9) is that presented by a particularly large random term causing x to take a negative value precluded by the natural barrier inherent in the SDE itself. This potential inconsistency can be removed by working with the logarithm of x, which can legitimately take negative values. It is straightforward to construct the SDE satisfied by $y = \log x$; application of Ito's formula yields

$$\begin{aligned} dy &= \left(\frac{dy}{dx}\right)dx + \frac{1}{2}\left(\frac{d^2y}{dx^2}\right)(dx)^2 \\ &= \frac{1}{x}\left((\nu - x)dt + \sqrt{2x}\,dW\right) - \frac{1}{2}\frac{1}{x^2}\cdot 2x\,dt \\ &= \left(\frac{\nu - 1}{x} - 1\right)dt + \sqrt{\frac{2}{x}}\,dW \\ &= ((\nu - 1)\exp(-y) - 1)dt + \sqrt{2}\exp\left(\frac{-y}{2}\right)dW. \end{aligned} \tag{11.70}$$

While this identification of the appropriate SDE is sufficient for us to construct an Euler type algorithm, essentially by inspection, we might wish to include terms of higher order in the time step, much as in the direct simulation of the gamma variate itself. To calculate the required higher order statistics of the increment in the log gamma process we again make use of the analytic form (11.59) of the propagator of the gamma Markov process. The generating function approach adopted earlier has to be modified slightly. An appropriate moment generating function can be constructed as follows:

$$\exp(s(\log(x) - \log(x_0))) = \left(\frac{x}{x_0}\right)^s$$

$$\left\langle \left(\frac{x}{x_0}\right)^s \right\rangle = 1 + s\langle(\log(x) - \log(x_0))\rangle + \frac{s^2}{2}\left\langle(\log(x) - \log(x_0))^2\right\rangle \cdots.$$

$$\tag{11.71}$$

We evaluate this function (essentially the Mellin transform of the propagator (11.59)) as

$$\left\langle \left(\frac{x}{x_0}\right)^s \right\rangle = \int_0^\infty (x/x_0)^s P(x,t|x_0)\,dx$$

$$= \frac{1}{1-\eta}\frac{1}{(x_0\eta)^{(\nu-1)/2}}\exp\left(-\frac{x_0\eta}{1-\eta}\right)\frac{1}{x_0^s}\int_0^\infty x^{s+(\nu-1)/2}$$

$$\times \exp\left(-\frac{x}{1-\eta}\right) I_{\nu-1}\left(\frac{2\sqrt{xx_0\eta}}{1-\eta}\right) dx. \tag{11.72}$$

The integration over the modified Bessel function can be effected analytically (eqn 11.4.28 in Abramowitz and Stegun 1970) to give us

$$\left\langle \left(\frac{x}{x_0}\right)^s \right\rangle = \frac{(1-\eta)^s}{x_0^s}\exp\left(-\frac{\eta x_0}{1-\eta}\right)\frac{\Gamma(\nu+s)}{\Gamma(\nu)}{}_1F_1\left(\nu+s;\nu;\frac{x_0\eta}{1-\eta}\right). \tag{11.73}$$

Here we have encountered the confluent hypergeometric function ${}_1F_1$ (Chapter 13 *ibid.*). In the construction of the simulation algorithm we focus attention on the short time behaviour of this function, where $\eta \to 1$. Thus the standard series representation of the confluent hypergeometric function (eqn 13.1.2 *ibid.*) is not particularly useful in this context. Fortunately, (11.73) can be recast using the asymptotic representation

$${}_1F_1(a:b:z) \sim \frac{\Gamma(b)\exp(z)}{\Gamma(a)z^{b-a}}{}_2F_0\left(b-a, 1-a; z^{-1}\right); \quad z \to \infty; \quad \Re(z) > 0 \tag{11.74}$$

developed by Copson (1970). The term ${}_2F_0$ is the generalized hypergeometric function, whose formal series expansion is

$${}_2F_0(a,b;q) = \frac{1}{\Gamma(a)\Gamma(b)}\sum_{n=0}^\infty \frac{\Gamma(a+n)\Gamma(b+n)}{n!}q^n. \tag{11.75}$$

Using this we find that

$$\left\langle \left(\frac{x}{x_0}\right)^s \right\rangle \sim \eta^s {}_2F_0\left(-s, 1-\nu-s; \frac{1-\eta}{x_0\eta}\right), \quad \eta \to 1. \tag{11.76}$$

Finally, we introduce the exponentially decaying correlation function

$$\eta = \exp(-\mathcal{A}\Delta t). \tag{11.77}$$

We now make an expansion in both s and Δt from which we can extract the mean and mean square values of the increment in the logarithm. Thus we have

$$\left\langle \left(\frac{x}{x_0}\right)^s \right\rangle = 1 + s\left\{\left(\frac{\nu-1}{x_0} - 1\right)\mathcal{A}\Delta t + \left(\frac{3\nu - \nu^2 - 2}{x_0^2} + \frac{(\nu-1)}{x_0}\right)(\mathcal{A}\Delta t)^2 \cdots\right\}$$
$$+ s^2\left\{\frac{\mathcal{A}\Delta t}{x_0} + \left(\frac{\nu^2 - 5\nu + 5}{2x_0^2} + \frac{3 - 2\nu}{2x_0} + \frac{1}{2}\right)(\mathcal{A}\Delta t)^2 \cdots\right\} + \cdots. \tag{11.78}$$

From this we can read off

$$\langle (\log(x) - \log(x_0)) \rangle = \left(\frac{\nu-1}{x_0} - 1\right)\mathcal{A}\Delta t + \left(\frac{3\nu - \nu^2 - 2}{x_0^2} + \frac{(\nu-1)}{x_0}\right)(\mathcal{A}\Delta t)^2$$

$$\left\langle (\log(x) - \log(x_0))^2 \right\rangle = 2\frac{\mathcal{A}\Delta t}{x_0} + \left(\frac{\nu^2 - 5\nu + 5}{x_0^2} + \frac{3 - 2\nu}{x_0} + 1\right)(\mathcal{A}\Delta t)^2 \tag{11.79}$$

and calculate the variance of the logarithmic increment as

$$\left\langle (\log(x) - \log(x_0))^2 \right\rangle - \langle (\log(x) - \log(x_0)) \rangle^2 = 2\frac{\mathcal{A}\Delta t}{x_0} + \frac{4 - 3\nu + x_0}{x_0^2}(\mathcal{A}\Delta t)^2. \tag{11.80}$$

The linear terms in time in the mean and variance of the increment are those that one would expect from (11.70) and might incorporate into an Euler algorithm. The quadratic correction terms can be accounted for just as was done in (11.69) for the case of the gamma, as opposed to the log-gamma, process. In a final step we regenerate the gamma process by exponentiation. Thus, the log-based algorithm can be summarized as follows:

$$\begin{cases} y(n) = y(n-1) + ((\nu - 1)\exp(-y(n-1)) - 1)\mathcal{A}\Delta t \\ \quad + \begin{pmatrix} (3\nu - \nu^2 - 2)\exp(-2y(n-1)) \\ + (\nu - 1)\exp(-y(n-1)) \end{pmatrix}(\mathcal{A}\Delta t)^2 \\ \quad + \left(\sqrt{2\mathcal{A}\Delta t \exp(-y(n-1)) + \begin{pmatrix}(4-3\nu)\exp(-2y(n-1)) \\ + \exp(-y(n-1))\end{pmatrix}(\mathcal{A}\Delta t)^2}\right)g(n) \\ x(n) = \exp(y(n)). \end{cases} \tag{11.81}$$

The final component of the clutter simulation within the compound model is the generation of a gamma process, described by a SDE. Simple Euler algorithms are almost invariably used to effect such simulations. Algorithms of higher order accuracy have been discussed in the literature; their generality unfortunately renders them rather forbidding. However, in the case of the gamma process, significant analytic progress has been made and simple algorithms, of an accuracy equivalent to that of these more general methods, have been constructed for the generation of the gamma variate and its logarithm.

12

EXPERIMENTAL TESTS

In this chapter we study two independent examples of data gathered from electromagnetic scattering from random media. The first example involves scattering from a uniform randomizing optical phase screen, which is applied as a calibration tool for the intensity volatility in our diffusive model of K-distributed noise. The second example consists of ocean surface radar scattering data that contains K-distributed sea clutter, a multitude of 'clutter spike' anomalies, and a genuine target for which our methodology is used to isolate the anomaly with respect to range.

In both cases we shall study the electromagnetic scattered amplitude stochastic processes and compare the empirically observed volatilities with the theoretical predictions from our stochastic model. In the absence of anomalies we expect a strong correlation between these two quantities, via the standard statistical correlation measure

$$c(X,Y) = \frac{\langle (X - \bar{X})(Y - \bar{Y}) \rangle}{(\text{Var} X \, \text{Var} Y)^{1/2}}. \tag{12.1}$$

Observe that this standard measure, which satisfies $-1 \leq c(X,Y) \leq 1$, is zero for independent variables, unity for 'perfectly correlated' variables, and is invariant under (positive) affine transformations of the random variables $X \mapsto \lambda X + \mu$, $\lambda > 0$.

Since the data were supplied in discrete time, an important issue arises concerning the instantaneous observability of the stochastic volatility explained in Lemma 5.4. Although the relations (5.24), (5.25) are exact within the framework of the Ito calculus, these differential relations are idealizations of the incremental properties that can be physically measured. In practice, physical observations occur at a discrete set of times and, in accordance with the fact that the physical processes exist in *continuous* time, the significance of the Ito calculus is that sample times remain unspecified. For discrete time sampling of an Ito process q_t, satisfying the stochastic differential equation (SDE) $dq_t = \mu_t dt + \sigma_t dW_t$, we may write

$$\delta q_t = \mu_t \delta t + \sigma_t n_t \delta t^{1/2} + o(\delta t), \tag{12.2}$$

where n_t is drawn from a normal distribution with zero mean and unit variance $\mathcal{N}(0,1)$, so $\langle n_t^2 \rangle = 1$. It follows that $\delta q_t^2 = \sigma_t^2 n_t^2 \delta t + o(\delta t)$. Given the Ito interpretation of the (discrete) SDE (12.2), the independent increments property of W_t implies that σ_t and $\delta W_t = n_t \delta t^{1/2}$ are independent (this independence also holds if σ_t is non-Markov). Therefore, taking expectations $\langle \cdot \rangle$, at fixed t with respect to the random variable n_t, we find[33]

$$\sigma_t^2 = \langle \delta q_t^2 \rangle / \delta t + o(1). \tag{12.3}$$

If q_t is a diffusion then $\sigma_t = \sigma(t, q_t)$, so that $\delta q_t^2 = \sigma(t, q_t)^2 n_t^2 \delta t + o(\delta t)$. Strictly, the expectation $\langle \cdot \rangle$ above should be interpreted as an *ensemble* average, for a given t, q_t. Nevertheless, since the sample paths of q_t are continuous with unit probability, a local time approximation to $\langle \delta q_t^2 \rangle$ is obtained by smoothing the raw δq_t^2 values by averaging over a time window $[t-\Delta, t+\Delta]$. This approximation may be improved upon (O'Loghlen 2001) by ordering the q_t values before smoothing the time series for δq_t^2. More precisely, we can permute the discrete sample times $\{t_i\} \mapsto \{p(t_i)\}$ such that $\{q_{p(t_i)}\}$ is an ordered set, then evaluate the expectation $\langle \delta q_{p(t_i)}^2 \rangle$ by smoothing, and apply the inverse permutation to the result. This method provides an improved empirical estimate of dq_t^2/dt, to be compared with the theoretical prediction $\sigma(t, q_t)$, the latter of which is instantaneously deducible from the instantaneous observed values $\{q_t\}$.

12.1 Scattering at optical wavelength

The dynamical framework of Chapter 8 applies equally well to forward-and back-scattered radiation. Moreover, it is an exact description for all parts of the electromagnetic spectrum, the carrier frequency being represented by the constant $\mathcal{B}$. In this section we shall consider an optical forward scattering experiment. The same type of analysis would apply to many situations of optical scattering, e.g. propagation of laser light through turbulent atmospheric conditions, and in optical instrumentation in theoretical astronomy.

12.1.1 *Random phase screen*

The data consisted of a complex-valued time series (sampled at 10 kHz) for the I, Q components of the electromagnetic field (forward) scattered from a randomizing phase screen. The details of the experimental situation are described in Ridley *et al.* (2002) and the resulting intensity data were observed to be very closely K-distributed, as shown in Fig. 12.1. The squared volatility function ψ of (5.11) was assumed to have power law behaviour with respect to the intensity (cf. the discussion following (5.19), (5.20) above) and to deduce the appropriate power a comparison was made of $\langle |\delta z_t|^p \rangle$ and z_t using the smoothing technique above to calculate the expectation $\langle \cdot \rangle$. The results are shown in Fig. 12.2 and indicate a strong correlation between these quantities according to the measure (12.1), with a maximum occurring at $p \approx 2$. A comparison of the time series for this value of p is shown in Fig. 12.3. The analysis demonstrates that the squared

---

[33]We assume that the data are sampled at a sufficiently high frequency for the $o(1)$ term in (12.3) to be negligible so that $\delta t^{1/2} \gg \delta t$ and the δW_t contribution to (12.2) dominates the drift over a sample period, as evidenced, for example, in Fig. 12.3.

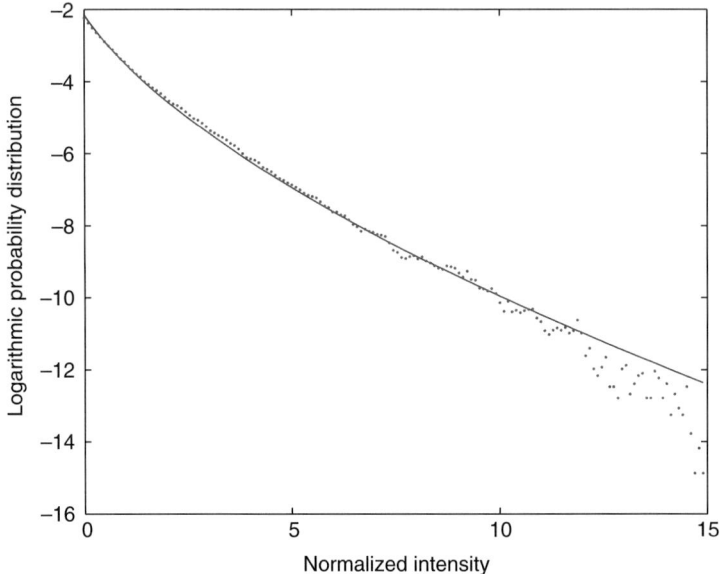

FIG. 12.1. Random phase screen data, the logarithmic probability distribution for the normalized forward-scattered intensity, showing comparison of the nearest fit theoretical K-distribution (solid line) and the experimentally observed histogram. (The discrepancy in the tail of the K-distribution is well known and a more refined model would be necessary to model such extreme intensities.)

volatility function ψ is to a close approximation equal to the intensity, which accords with the theoretical discussion of Section 8.1.

Remarks. *On the suppression of fluctuating terms.* Observe that the time series behaviour for this stochastic dynamics is radically different from what would be obtained from suppressing the fluctuating term in the underlying intensity SDE. Since the drift and volatility coefficients are functions of the state (the intensity itself), an abscissa of constant intensity will intersect such (deterministic) path at most once; however, this geometrical construct yields multiple crossings for the fluctuating time series, as readily apparent from Fig. 12.3.

12.2 Scattering at radar wavelength

In this section we provide a detailed experimental study of scattering at radar wavelength for a case of sea surface scattering. We examine the nature of the amplitude fluctuations in light of our theoretical results, and demonstrate a close agreement between theory and experiment. Our results also enable construction

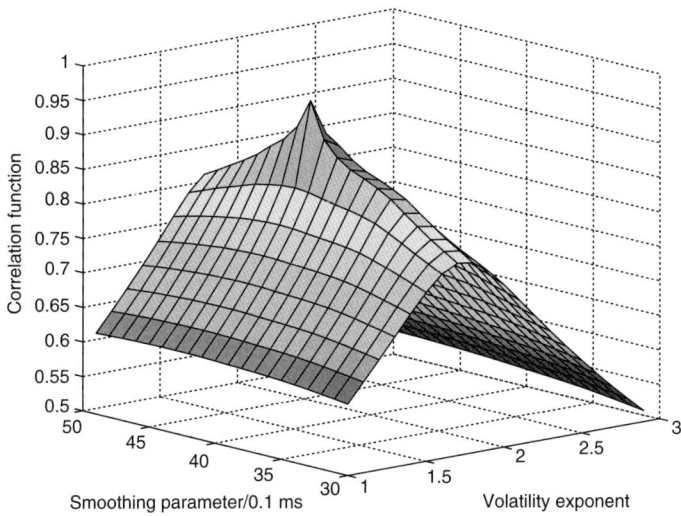

FIG. 12.2. Correlation of z_t and $|\mathrm{d}z_t|^p$ for random phase screen data. The values $p \approx 2$, smoothing window parameter $\Delta \approx 40$ give an optimal correlation of approximately 95%. For visualization purposes, the correlation function shown is $M - (M - c)^{1/2}$ where $M = \sup(c)$. The surface c is approximately flat near its maximum (i.e. the absolute value of its derivatives are close to zero). This is a desirable property from the point of view of the sensitivity of the technique to the precise values chosen for the volatility exponent p and smoothing parameter Δ, which are inevitably subject to empirical errors.

of an anomaly detection mechanism, which is applied to the sea surface scattering data to isolate a body in space to a high degree of accuracy.

12.2.1 Coherent sea clutter

The data consists of K-distributed radar returns from a region of the ocean surface, whose I, Q components (e.g. Helstrom 1960) constitute a complex-valued two-dimensional space-time array $\Psi(t, x) = I(t, x) + iQ(t, x)$, with pulse repetition frequency (PRF) 1kHz, over a range of 300 m at a resolution 0.3 m, and within which a tethered target is situated whose location was known. The intensity data are shown in Fig. 12.4. The data were monochromatic, so that persistent temporal correlation due to the illuminated physical objects can occur over appreciable timescales, and there is no decorrelation arising from component frequencies of the illuminating beam. In accordance with the findings from the anomaly-free K-distributed data of Section 12.1, we adopt the behaviour $\sigma^2_{(z)} \sim z$, i.e. $p \approx 2$ for the stochastic volatility of the intensity process. The view is therefore that the appropriate volatility exponent is a consequence of the physics of the scattering process, which is common to both sets of K-distributed data. A validation of this choice of calibration shall be a predominance of high correlation between the various predicted and observed volatilities under this

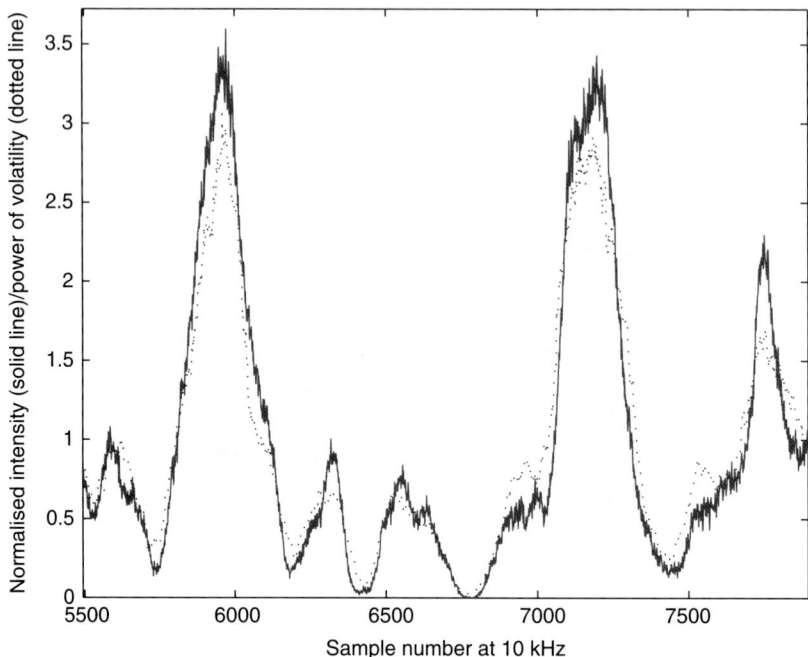

FIG. 12.3. Random phase screen data – comparison between raw intensity and $|dz|^p$ (both normalized by their respective means over the time shown). The figure shows a strong correlation for $p \approx 2$.

assumption, as evident in Fig. 12.6. The methodology of Section 12.1 involving only the intensity process, when applied to the current set of data revealed a preponderance of 'clutter spike' anomalies with respect to range. As a result of this, our analysis was enhanced by studying the stochastic volatility behaviour of the complex-valued amplitude Ψ directly. It is the inclusion of the phase information contained in Ψ that is the important feature in the avoidance of false alarms caused by 'clutter spike' anomalies (Luttrell 2001), and which contains important correlation information between the I and Q components. Accordingly, the analysis of Ψ exploits all the physical degrees of freedom in the electromagnetic field, which is desirable from the point of view of discrimination. (From a computational point of view the analysis involving Ψ_t is only of the order a factor of two more intensive compared to the corresponding analysis of the phase screen data which involved only the intensity z_t.)

For a given range we express the complex-valued amplitude in polar form $\Psi_t = R_t \exp(i\theta_t)$, so that $R_t = \sqrt{z_t}$. In order to calculate $d\Psi_t$, we apply Ito's formula to the modulus amplitude R_t, which yields $dR_t = dz_t/2z_t^{1/2} - dz_t^2/8z_t^{3/2}$. Writing $dz_t = \mathcal{B}b_{(z)}dt + (2\mathcal{B})^{1/2}\sigma_{(z)}dW_t^{(R)}$ so that $dz_t^2 = 2\mathcal{B}\sigma_{(z)}^2 dt$, we deduce that

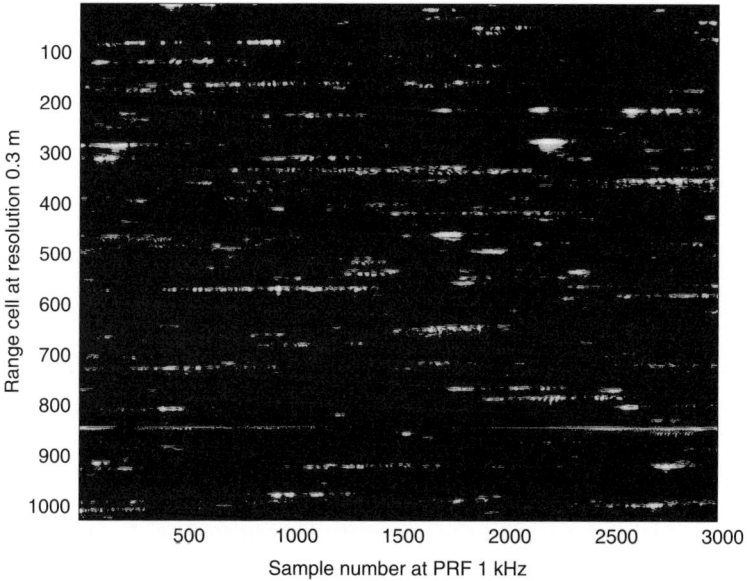

FIG. 12.4. Sea clutter space-time intensity.

$$dR_t = \frac{\mathcal{B}}{4R^3}\left(2R^2 b_{(z)} - \sigma_{(z)}^2\right) dt - \left(\frac{\mathcal{B}}{2R^2}\right)^{1/2} \sigma_{(z)} dW_t^{(R)}. \qquad (12.4)$$

Ito's formula yields the identity $d\exp(i\theta_t) = \exp(i\theta_t)[id\theta_t - \frac{1}{2}d\theta_t^2]$ and thus by the Ito product rule $d(X_t Y_t) \equiv X_t dY_t + Y_t dX_t + dX_t \, dY_t$ (see e.g. Oksendal 1998) we deduce

$$d\Psi_t = \exp(i\theta_t)\left[iR_t d\theta_t + dR_t + idR_t \, d\theta_t - \frac{1}{2} R_t d\theta_t^2\right]. \qquad (12.5)$$

This identity yields the following geometric expressions for the squared volatilities of the amplitude process Ψ_t,

$$d\Psi_t^2 = \exp(2i\theta_t)\left[dR_t^2 - R_t^2 d\theta_t^2 + 2iR_t dR_t \, d\theta_t\right], \qquad (12.6)$$

$$|d\Psi_t|^2 = dR_t^2 + R_t^2 d\theta_t^2 \qquad (12.7)$$

The phase θ_t is taken to be a stochastic quantity driven by an *independent* Wiener process $W_t^{(\theta)}$ and which obeys the SDE $d\theta_t = \omega dt + \sigma_{(\theta)} dW_t^{(\theta)}$. The independence of $W_t^{(R)}$, $W_t^{(\theta)}$ renders the $dR_t \, d\theta_t$ term in (12.6) zero (see e.g. Karatzas and Shreve 1988). An explicit derivation of this independence in the Wiener increments in R, θ from a random walk model is provided in Field and Tough (2003b). Thus we obtain the following result.[34]

---

[34] Observe that the absolute value and square operations on $d\Psi_t$ do not commute with one another, i.e. $|d\Psi_t^2| \neq |d\Psi_t|^2$.

Proposition 12.1 *The squared volatilities of the amplitude process Ψ_t are given in terms of the intensity and phase squared volatilities by*

$$\mathrm{d}\Psi_t^2 = \exp(2i\theta_t) \left(\frac{\mathcal{B}\sigma_{(z)}^2}{2z_t} - z_t \sigma_{(\theta)}^2 \right) \mathrm{d}t, \tag{12.8}$$

$$|\mathrm{d}\Psi_t|^2 = \left(\frac{\mathcal{B}\sigma_{(z)}^2}{2z_t} + z_t \sigma_{(\theta)}^2 \right) \mathrm{d}t. \tag{12.9}$$

Observe that Proposition 12.1 does not involve drift quantities and only the instantaneous *volatilities* in the intensity and phase feature explicitly in (12.8), (12.9).

In the pure Rayleigh case, the I, Q processes can be described by the pair of (uncoupled) Ornstein–Uhlenbeck processes (6.3), (6.4) from which we deduce (e.g. via Ito's formula applied to $\arctan(I/Q)$) that $\sigma_{(\theta)}^2 = 2/z$ thus exhibiting singular behaviour at zeroes of the intensity. This suggests that such singular behaviour should occur also in the K-distributed case which constitutes a Rayleigh process with a modulated cross-section. The phase volatility can be calculated from the normalized amplitude process via the identity $|\mathrm{d}\exp(i\theta_t)|^2 = \mathrm{d}\theta_t^2$, which can be applied directly to the data and avoids difficulties arising from the discontinuity in θ_t at 2π. A comparison of the intensity and (squared) phase volatility is shown in Fig. 12.5, which reveals singular behaviour in $\sigma_{(\theta)}$ near zeroes of the intensity, characteristic of the Ornstein–Uhlenbeck (Rayleigh) process, and approximate constant behaviour when the intensity lies above some threshold.

In respect of the singular behaviour at zeroes of the intensity, our results are consistent with studies of the phase derivative, for a *differentiable* amplitude process, reported in Jakeman et al. (2001). In contrast, however, in the differentiable case it is the *intensity*, rather than amplitude-weighted phase derivative that has minimal variance. This distinction between the differentiable and diffusion models for the amplitude Ψ_t should enable one to select which type of model is appropriate in different physical situations (see *orig.* §IIC in Field and Tough 2003b for a detailed theoretical account of the phase behaviour in the non-Gaussian case).

From a physical point of view the behaviour of $\sigma_{(\theta)}^2$ is anticipated as follows. The amplitude $\Psi = I + iQ$ is generated from (some component of) the complex-valued electric field $\mathcal{E}$. One therefore expects a contribution to $\sigma_{(\theta)}$ that varies inversely with R, since $|\delta\mathcal{E}_{(\theta)}| \sim R\delta\theta$ and $\langle|\delta\mathcal{E}_{(\theta)}|\rangle$ contains an 'intensive' part that is independent of the total field intensity $\mathcal{E}_{\mathrm{tot}}^2$. This observation is consistent with the Ornstein–Uhlenbeck description of the Rayleigh process given above. In the random walk model (Jakeman and Tough 1988) with $\mathcal{E}_{\mathrm{tot}} \sim \sum_{n=1}^{N} \exp(i\varphi_n)$, there is a fluctuation term that relates only to the random phases φ_n and which is reflected in $W^{(2)}$ of the intensity SDE (5.20). Simultaneously, in the case of a time varying scattering cross-section, there exist fluctuations in θ_t induced by an

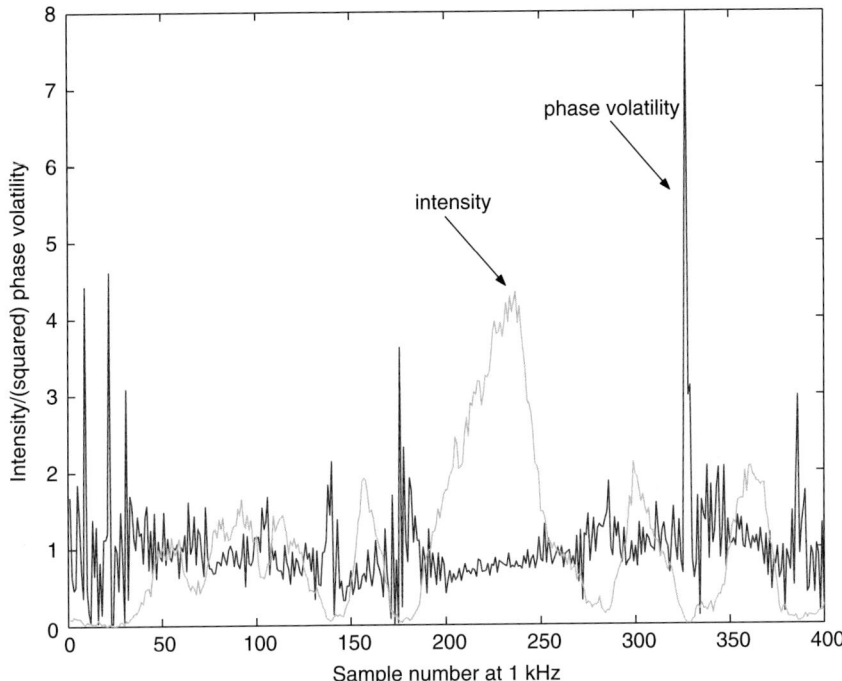

FIG. 12.5. Comparison of intensity and squared phase volatility.

'extensive' contribution to $\langle |\delta\mathcal{E}_{(\theta)}|^2\rangle$ that scales with the scattering cross-section $\langle |\mathcal{E}_{\text{tot}}|^2\rangle$, i.e. the number of steps N in the random walk model of Jakeman and Tough (1988). This feature is consistent with the appearance of a square root volatility in the SDE for the scattering cross-section (5.19). For a Rayleigh (Gaussian speckle) process the cross-section x_t is constant, and so the 'extensive' contribution to $|\delta\mathcal{E}_{(\theta)}|$ does not arise.

Assuming that $dW^{(R,\theta)}$ are independent, i.e. $dW_t^{(R)}dW_t^{(\theta)} = 0$, it follows from the coordinate transformations $I = R\cos\theta$, $Q = R\sin\theta$ that $dW_t^{(I)}dW_t^{(Q)}$ vanishes if and only if $\sigma_{(R)} = R\sigma_{(\theta)}$, or equivalently $\sigma_{(z)}^2 = 4z^2\sigma_{(\theta)}^2$. This special relation holds in the Rayleigh case for which the Wiener increments in I, Q are independent. In more general circumstances, however, we expect a departure from this relation, so that in the K-distributed case the fluctuations in the scattering cross-section induce a correlation between the increments dI_t, dQ_t.

Explicit derivations of these various properties of the amplitude process in the K-distributed case from the random walk model are described in Chapter 8. These calculations reveal that $\sigma_{(z)}^2$ contains a term proportional to the constant $\mathcal{A}$ that is quadratic in z, and a linear term proportional to $\mathcal{B}$. This explains the observed behaviour, for $\mathcal{A} \ll \mathcal{B}$, of the volatility exponent $p \approx 2$ in the analysis of the intensity data, and yields a linear term in $|d\Psi_t|^2$ according to

(12.9). The analysis of the phase shows that $\sigma^2_{(\theta)}$ is proportional to the constant $\mathcal{B}$ and varies inversely with z and linearly with x in accordance with the general scaling arguments adduced above. Consequently for $\mathcal{A} \neq 0$, i.e. the K-distributed case, the Wiener increments in the I, Q components of the field are correlated. The special relation above for independence of the quadrature components is recovered in the Rayleigh case for which $\mathcal{A} = 0$.

12.2.2 Anomaly detection

As a consequence of Proposition 12.1 and the invariance of $c(X, Y)$ under linear transformations, therefore, we obtain the following result.

Corollary 12.2 *For a K-distributed[35] process described by (5.19), (5.20) the instantaneous correlation between the observed and theoretical $|d\Psi_t|$ squared volatilities reduces to*

$$c_\Psi = c(|d\Psi_t|^2, z_t). \tag{12.10}$$

The results of applying this correlation measure at each range cell are shown in Fig. 12.6, which reveals high correlation at all ranges except the target location. The persistence of this local spatial de-correlation over time is shown in Fig. 12.7. Using these correlation results, plots were generated of the time series for the theoretical and observed $|d\Psi|^2$ squared volatilities at both the maximum and minimum correlation ranges as discerned from Fig. 12.6. This was carried out in both real and permuted time, as explained above. The results shown in Figs. 12.8–12.11 illustrate the instantaneous nature of the correlation, and how this is destroyed at the target location.

The plots of the inverse predictive functions shown in Figs. 12.9 and 12.11 in fact provide the approximate cumulative distribution functions (c.d.f.) for the respective random variables, when the vertical axis of the inverse function is re-scaled to the interval $[0, 1]$. This feature would be useful in obtaining a measure of discrimination between the corresponding distributions, since the Fisher information geodesic distance between two distributions p_i can be approximated directly in terms of the discrete c.d.f.s c_i as $\cos\theta_{ij} = \sum_\alpha [\delta c_i(\alpha) \delta c_j(\alpha)]^{1/2}$. In the present approach, however, we assume that, in the clutter domain, the underlying distributions are *identical* and apply the statistical correlation measure (12.1) to the distinct random variables. It may however be possible to exploit the Fisher information discrimination measure through a corresponding SDE description for the target (cf. Wootters 1981).

The approach to anomaly detection described in the present section should be compared with Haykin (1999) in which real empirical data are used to build the predictive sea clutter model.

It is of some (theoretical) interest at this point to consider the possibility of an anomaly detection scheme based on the same type of principle of comparison

---

[35]This result would not hold for the pure Rayleigh process because in that case $\sigma^2_{(z)} \propto z$, $\sigma^2_{(\theta)} \propto 1/z$ and the right-hand side of (12.9) is then constant.

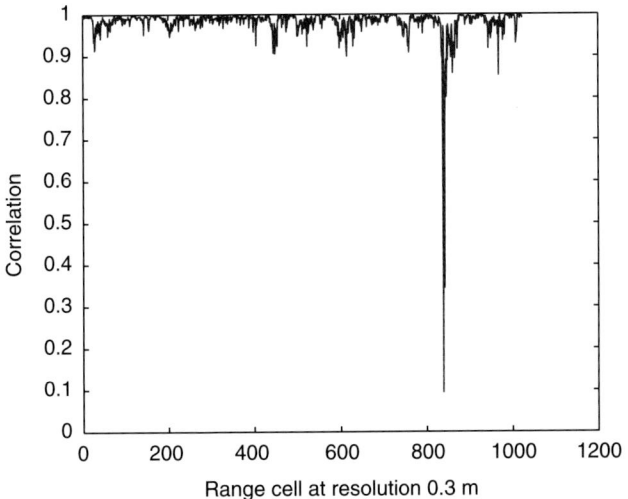

FIG. 12.6. Correlation of $|d\Psi|^2$ and σ_Ψ^2 according to Corollary 12.2 versus range. Mean correlation = 98.11%. Maximum correlation 99.94% occurring at range cell 725. Minimum correlation 9.48% occurring at range cell 839. A threshold of 80% isolates the anomaly to within 4 range cells, i.e. approximately 1 m. (The signal is sampled for 1 s at PRF 1kHz over a range 360 m. The smoothing parameter Δ is 50 samples.)

between observed and theoretical squared volatilities, but that does not involve the scattering cross-section x_t explicitly. Indeed, such a scheme can be readily constructed, as follows (albeit up to second order in the time differential). First recall the squared differential relations for the amplitude and its modulus:

$$\begin{cases} |d\Psi_t|^2 = (\mathcal{B}x_t + \mathcal{A}z_t/2x_t)dt, \\ dR_t^2 = \frac{1}{2}(\mathcal{B}x_t + \mathcal{A}z_t/x_t)dt. \end{cases} \quad (12.11)$$

Taking linear combinations of these yields

$$\begin{cases} |d\Psi_t|^2 - dR_t^2 = \frac{1}{2}\mathcal{B}x_t dt \\ 2dR_t^2 - |d\Psi_t|^2 = \frac{\mathcal{A}z_t}{2x_t}dt \end{cases} \quad (12.12)$$

whose product gives the quadratic time differential relation

$$\left(|d\Psi_t|^2 - dR_t^2\right)\left(2dR_t^2 - |d\Psi_t|^2\right) = \frac{1}{4}\mathcal{A}\mathcal{B}z_t dt^2. \quad (12.13)$$

In principle, this general relation facilitates an alternative stochastic volatility based anomaly detection, without concern for the properties of x_t which is notably absent in (12.13).[36]

---

[36]The author is grateful to Dr. Tao (Stephen) Feng for pointing out the existence of such a scheme.

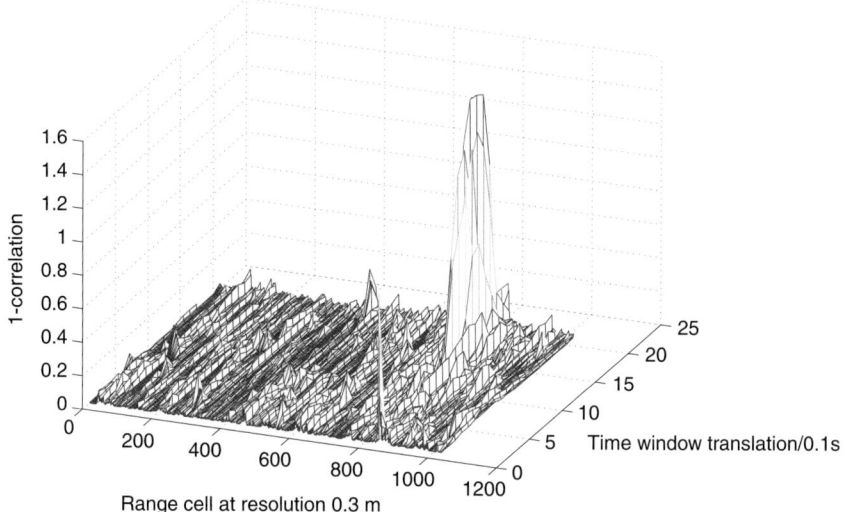

FIG. 12.7. 3-dimensional plot showing range and time horizontally and $1 - c(|\mathrm{d}\Psi_t|^2, z_t)$ vertically; each point along the time axis corresponds to a windowed sample of 1 s, thus covering a total time duration of 3 s of data. The figure illustrates the persistence of the anomaly detection mechanism over time – the object of interest disappears out of view for a short period while it is submerged due to bulk wave motion, and then re-surfaces.

We close this chapter with some concluding remarks on the experimental aspects of our scattering models. Our stochastic treatment of the electromagnetic scattering from random media reveals a new technique for anomaly detection, which is based on the instantaneous observability of the stochastic volatility of the complex-valued amplitude process. The emphasis on the stochastic volatility constitutes a fundamental change in viewpoint as to the choice of significant parameters, and lies in contrast to previous thinking which has focussed on the dynamics contained in the drift. In this respect, there is a similarity between our approach to the scattering problem and the mathematical theory of option pricing for which the absence of an ensemble average leads to an arbitrage-free pricing mechanism that is drift independent (cf. Appendix D).

In the analysis of the radar data of Section 12.2, an essential ingredient has been the behaviour of the complex-valued amplitude process, and how the form of its volatility function induces correlation properties between the I, Q components of the amplitude. In the K-distributed case, these correlations arise from fluctuations in the scattering cross-section. Such correlations are correspondingly absent in the pure Rayleigh (Gaussian speckle) process, for which there exists a description for I, Q in terms of a pair of uncoupled Ornstein–Uhlenbeck processes.

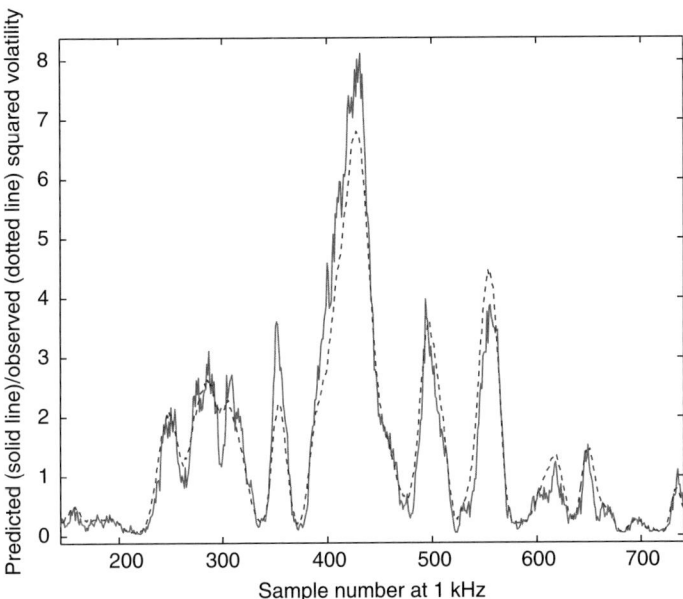

FIG. 12.8. Comparison of predicted and measured (smoothed) squared volatility for the process $|d\Psi_t|$. The range cell was chosen to give *maximum* correlation.

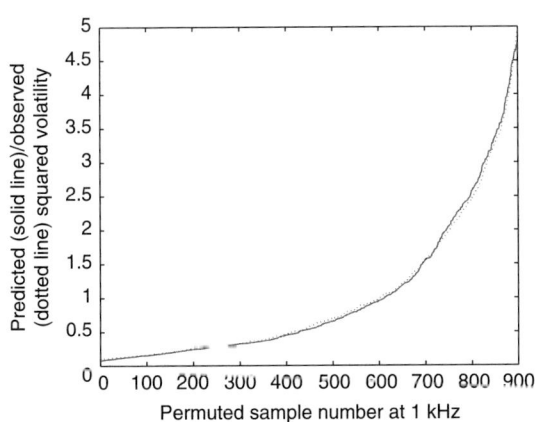

FIG. 12.9. Permuted-time representation of predicted and observed (smoothed) squared volatility for the process $|d\Psi_t|$. The predicted values are permuted in ascending order; the required permutation is then applied also to the observed values, which are subsequently smoothed. The range cell was chosen as in the previous figure.

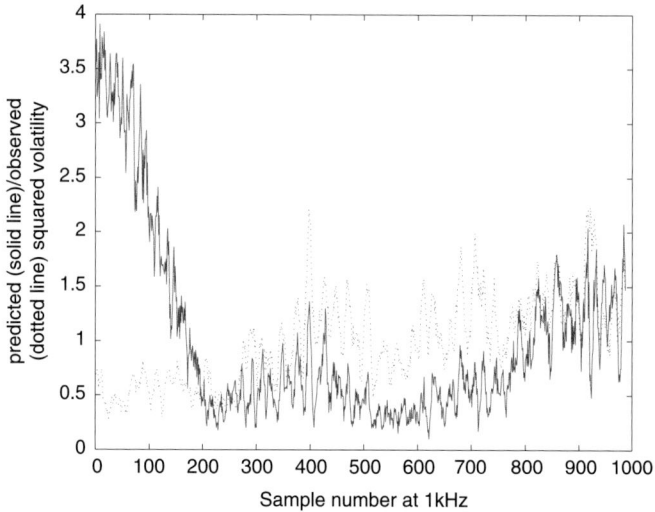

FIG. 12.10. Comparison of predicted and observed volatilities; as in Fig. 12.8 except with range cell chosen at the location of the anomaly where the correlation is *minimized*.

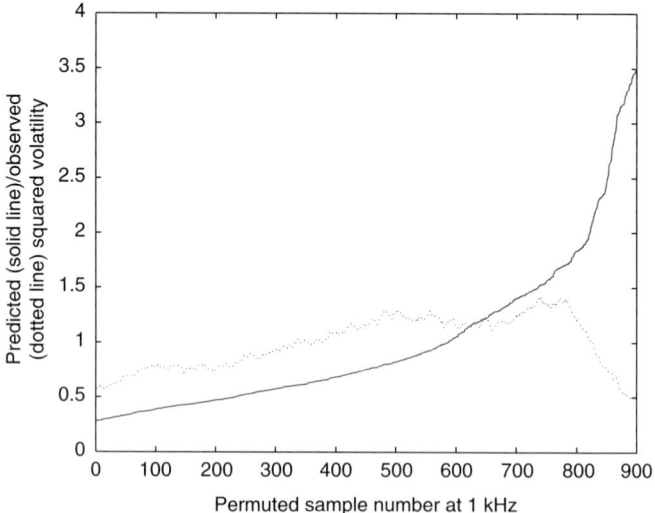

FIG. 12.11. Comparison of predicted and observed volatilities; as in Fig. 12.9 except with range cell chosen at the location of the anomaly where the correlation is minimized.

In furtherance of this study one hopes to exploit the freedom in the volatility functions in the classification of K-distributed processes provided in Chapter 5, thus exploring different calibrations for $\sigma_{(z)}$ and the underlying phase volatility $\sigma_{(\theta)}$ from a simulation or data-driven point of view. Our analysis could be extended to include non-Markovian models for the behaviour of x_t, z_t, Ψ_t via the introduction of extra variables into a larger number of coupled SDEs (cf. Tough 1987), and in addition, to explore other models for the scattering cross-section, which do not fall into the general category of Ito processes. The analysis of the discrete sampled data in this chapter, which integrates the underlying Ito SDEs to first order only, could be enhanced by higher order numerical integration schemes (Klauder and Petersen 1985) as described in Chapter 11. The aim of pursuing these various avenues would be to optimize the correlation between the predicted and observed volatilities in the predominant K-distributed domain. The techniques we have described could also likely be adapted to situations that occur in the modelling of financial time series with a view to detecting anomalous behaviour, and to other situations of propagation of acoustic waves through random media.

13
NON-LINEAR DYNAMICS OF SEA CLUTTER

We review experimental evidence for the non-linearity of sea clutter and the role of the z-parameter or Mann–Whitney rank-sum statistic in quantifying this non-linear behaviour in the context of a hybrid amplitude modulation (AM)/frequency modulation (FM) model for sea clutter, viewed as a cyclostationary process. An independent theoretical derivation of the stochastic dynamics of radar scattering in a sea clutter environment, in terms of a pair of coupled stochastic differential equations (SDEs) for the received envelope and radar cross-section (RCS), enables the identification of non-linearity in terms of the shape parameter for the RCS. We are led to conclude that, from both experimental and theoretical points of view, the dynamics of sea clutter are non-linear with a consistent degree of non-linearity that is determined by the sea state.[37]

Haykin et al. (2002) advocated a state-space formalism for the processing of radar signals in the presence of sea clutter (i.e. radar backscatter from an ocean surface). Such a model not only accounts for the temporal dimension of sea clutter in an explicit manner but also its statistical characterization. Basic to this formalism is whether the underlying dynamics of sea clutter are linear, or non-linear.

In the detailed experimental study reported in Haykin et al. (2002), it was also demonstrated that sea clutter is a non-linear dynamic process, with the degree of non-linearity increasing as the 'sea state' becomes higher. The conclusion reached on the non-linearity of sea clutter was based on two premises, using real-life data collected with an instrument-quality coherent radar system:

1. The characterization of sea clutter embodies two forms of continuous-wave modulation:

(i) amplitude modulation, which is linear, and
(ii) frequency modulation, which is non-linear

The latter phenomenon is responsible for the non-linearity of sea clutter.

2. The z-parameter, denoting the Mann–Whitney rank-sum statistic, is less than the special value -3, which is a strong indicator of non-linearity.

With regard to point 1, it is also noteworthy that in another study that focussed on the spectral characterization of sea clutter using the Loève transform (Haykin and Thomson 1998), it was discovered for the first time that sea

---

[37]The author acknowledges the input and collaboration of Prof. Simon Haykin with regard to the material on the z-parameter and the hybrid AM/FM model in this chapter.

clutter is a cyclostationary process. Cyclostationarity is ordinarily associated with modulation. But knowing that sea clutter is cyclostationary, it does not tell us the type of modulation involved in the characterization of its waveform.

In this section, we expand on the characterization of sea clutter as a non-linear dynamic process, using a principled theoretical approach. In particular, the approach is rooted in SDE theory. The issue of the dynamics of radar scattering in a sea clutter environment has been addressed in the literature independently from both theoretical and experimental points of view. Perhaps most notable in the former case, Field and Tough (2003a,b) develop a theoretical basis for the dynamics which is demonstrated to agree with experimental data to a remarkable degree of accuracy. In the latter case, Haykin et al. (2002) study experimental data to motivate a line of argument leading to the conclusion that sea clutter is inherently non-linear (and, indeed, possibly chaotic). Here we bring these two independent lines of development together in a consistent way in order to establish the non-linear nature of sea clutter from both physical and mathematical viewpoints. More precisely, the scattering dynamics can be derived from first principles in terms of a pair of SDEs for the received envelope and the radar cross-section (RCS) that feature a non-linear coupling and encode the statistical character of the sea state in terms of a certain 'shape parameter'. Examination of the differentiable parts in this system of SDEs reveals a corresponding 'noise-free skeleton' that is a non-linear vector process, with a degree of non-linearity dependent on the shape parameter in a manner consistent with that shown experimentally by Haykin and co-workers. This significant development affirms the case for the non-linear character of radar sea clutter.

The chapter is organized as follows. Section 13.1 provides a summary of the experimental study that led to the formulation of a hybrid AM/FM model, and the conclusion that sea clutter is a non-linear dynamic process. Section 13.2 summarizes the essential ingredients of SDE theory necessary for the basic interpretation of the SDE dynamics of radar sea clutter. We then apply this formalism to establish the non-linear character of the stochastic dynamics of the vector process consisting of the RCS and resultant back-scattered amplitude or 'received envelope'. This is achieved from first principles via an extended random walk model, as developed in Chapter 8. The extent of the non-linearity in the resulting SDE description is quantified in terms of a certain 'shape parameter' (the relative variance in the RCS, minus one) that encodes the sea state. We conclude with a discussion of the interplay between the two independent lines of enquiry that lead to the common conclusions concerning the non-linear character of radar sea clutter. We also indicate how our results may suggest which types of experiments to perform to further substantiate and enhance the theoretical framework, and discuss future prospects for the investigation of chaotic dynamics.

We refer the reader also to the recent book by Haykin (2006 ed.), where the experimental results are mentioned in the broader context of adaptive radar signal processing.

13.1 Hybrid AM/FM model of sea clutter

In an independent study reported in Gini and Greco (2001), sea clutter was viewed as a fast 'speckle' process multiplied by a 'texture' component that represents the slowly varying power level of the sea clutter signal; such a model is perceptually satisfying. The slow variation of the sea clutter power level was attributed to the large ocean waves passing through the observed ocean patch. The speckle was modeled as a stationary compound complex Gaussian process, and the texture was modelled as a harmonic process.

Inspired by the Gini–Greco model of sea clutter, Haykin and coworkers carried out an extensive physical study of sea clutter collected by the instrument quality coherent IPIX radar, where the radar data were recorded on the East Coast of Canada (Haykin et al. 2002). In that paper, it was demonstrated that AM and FM play important roles in the waveform description of sea clutter. The hybrid AM/FM model of sea clutter has been substantiated further in Greco and Gini (2007).

To explain the physical presence of modulation in sea clutter, we observe that when a large wave passes across a patch of the ocean surface, it will first accelerate and then decelerate the water's motion on the ocean surface. The continuous tilting of the ocean surface by the waves gives rise to amplitude modulation.

Moreover, the ocean wave will cause a cyclic motion of the instantaneous velocity of scatterers on the ocean surface, thereby giving rise to frequency modulation as another characteristic of the sea clutter waveform. When the mean velocity of the scatterers is high at a given instant of time, then the spectral spread (i.e., the bandwidth occupied by the frequency modulation) around that mean is correspondingly high, which is in perfect accordance with modulation theory.

It is well known that, unlike AM, FM is a non-linear process (Haykin 2001). Therefore, the presence of FM in the physical behaviour of the sea clutter waveform leads us to hypothesize that sea clutter is a non-linear dynamic process. To validate this hypothesis Haykin et al. use 78 different coherent radar data sets to compute the z-parameter, which denotes the Mann–Whitney rank-sum statistic (Siegel 1956). The results of this test are reproduced in Fig. 13.1, where the z-parameter is plotted against the spectral width modulation. A value of z less than -3 is considered to be a strong reason for rejecting the null hypothesis that the sea clutter data under test can be described by linearly correlated noise. In Fig. 13.1, we clearly see that the large majority of the experimental points lie below $z = -3$. Those points were in actual fact representative of high sea states. On the basis of these experimental results, Haykin et al. concluded that sea clutter is indeed a non-linear dynamic process, with the degree of non-linearity increasing with increasing sea state.

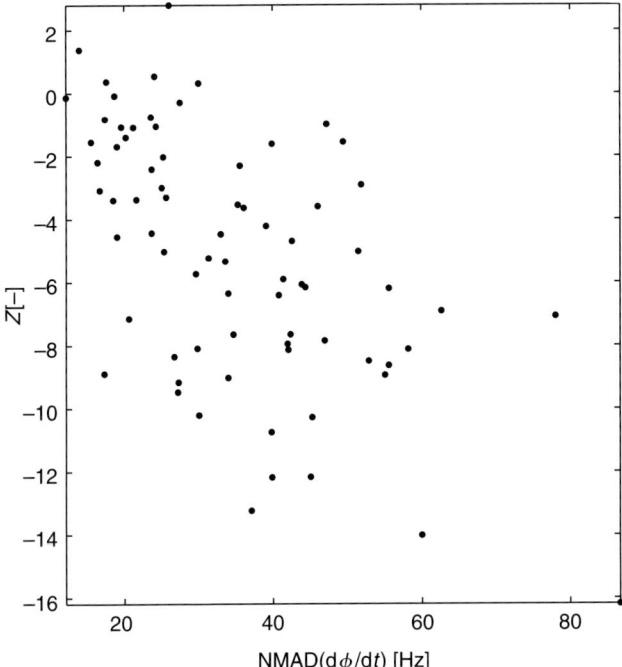

FIG. 13.1. z-parameter value versus Normalized Median Absolute Deviation NMAD ($\dot\phi$), computed for 78 data sets measured by the IPIX radar at various experimental conditions.

13.2 Non-linear dynamics from SDE theory

SDE theory has significant implications for statistical signal processing. It has recently proven successful in this context in the application to radar sea clutter (Field 2002; Field and Tough 2003a; Field and Haykin 2008). In a more general physical context, including optical propagation, the stochastic calculus has led to substantial new theoretical developments in the subject of electromagnetic scattering from random media (Field and Tough 2003b; Field 2005).

More recently, SDE techniques have been applied to wireless channel modelling (Feng *et al.* 2007) to include the effects of phase fluctuations in multi-path reception. The fact that similar techniques are applicable to both the radar backscattering and wireless propagation problems stems from the fact that each is multi-path in nature, with the only essential difference being that for radar the receiver and transmitter are co-located. This latter feature, however, does not affect the structure of the mathematical model used to describe the resulting amplitude signal.

Here we shall consider the RCS and received envelope processes to evolve according to the dynamics governed by a SDE. In the context of the RCS such dynamics arise from taking the continuum limit of a generic population dynamic model for the (discrete) number of component scatterers. For the scattered radiation the origin of the SDE dynamics lies in the behaviour of the component phases which are taken to evolve in time according to a Wiener process on a suitable (Rayleigh) timescale.

Thus we are able to represent the essential ingredients of the radar backscatter temporally, in the form of a continuous time stochastic process, say q_t, which evolves in time according to

$$\mathrm{d}q_t = b_t \mathrm{d}t + \sigma_t \mathrm{d}W_t. \tag{13.1}$$

Herein, b_t is a random process, and represents the ordinary time derivative of the process q_t in the case that σ vanishes. The quantity σ_t, on the other hand, is the amplitude of the noise or fluctuating part of q_t, in general a random process, and referred to as the 'stochastic volatility' of q_t. In the cases we study, it will become apparent that $b_t = b(t, q_t)$ and $\sigma_t = \sigma(t, q_t)$ for some specific functions b, σ, and accordingly the process q_t is called a 'diffusion'.

In contrast to the part of $\mathrm{d}q_t$ containing b_t, the σ_t term contributes an essential part to q_t that is not differentiable, in the ordinary sense that $\mathrm{d}/\mathrm{d}t$ is well-defined. Nevertheless, the (Ito) stochastic differential of q_t can be well defined.

In the engineering physics literature, one is perhaps more familiar with the Langevin equation for the time derivative

$$\frac{\mathrm{d}q}{\mathrm{d}t} = b_t + \sigma_t \Gamma_t \tag{13.2}$$

in which Γ_t is the familiar white noise process and Γ_t has the autocorrelation property $\langle \Gamma_t \Gamma_{t'} \rangle = \delta(t - t')$. For our purposes it will be sufficient to understand and interpret from the dynamical equations for the RCS and the received radar amplitude that, in a discrete-time setting,

$$\delta q_{t_i} = b_{t_i} \delta t + \sigma_{t_i} n_{t_i} \delta t^{1/2}, \tag{13.3}$$

where $\{t_i\}$ is a discrete set of observation times, $\delta t = t_{i+1} - t_i$, and $\{n_{t_i}\}$ are a collection of independent $\mathcal{N}(0,1)$ random variables. Then the above properties of q and its time derivative are evident (see Oksendal 1998 for a detailed rigourous account).

The essence of the approach taken is therefore to postulate the exact dynamics in continuous time, and then sample at a discrete set of times corresponding to the physical measurements. This procedure is inevitably more precise than an attempt at a model that is fundamentally discrete time in nature, since the physical observables are not quantized in time.

We shall assume the (dynamical extension of the) random walk model for the resultant back-scattered amplitude or 'received envelope'

$$\mathcal{E}_t^{(N)} = \sum_{j=1}^{N} \overbrace{a_j \exp\left[i\varphi_t^{(j)}\right]}^{s^{(j)}} \quad (13.4)$$

with (fluctuating) population size N, random phasor step $s^{(j)}$, and 'form factors' a_j. The key result of relevance to our discussion is obtained by taking the (Ito) stochastic differential of (13.4). This provides the following coupled stochastic dynamics of the RCS and scattered amplitude/received envelope Ψ_t.[38]

Proposition 13.1 *The dynamics of the RCS and received envelope for radar sea clutter, with shape parameter $\nu = \alpha - 1$, are given by the following set of non-linearly coupled SDEs:*

$$\mathrm{d}x_t = \mathcal{A}(\alpha - x_t)\mathrm{d}t + (2\mathcal{A}x_t)^{1/2}\mathrm{d}W_t^{(x)} \quad (13.5)$$

$$\frac{\mathrm{d}\Psi_t}{\Psi_t} = \left[\mathcal{A}\left(\frac{2(\alpha - x_t) - 1}{4x_t}\right) - \frac{1}{2}\mathcal{B}\right]\mathrm{d}t + \left(\frac{\mathcal{A}}{2x_t}\right)^{1/2}\mathrm{d}W_t^{(x)} + \frac{\mathcal{B}^{1/2}}{\gamma_t}\mathrm{d}\xi_t \quad (13.6)$$

in which γ_t is a unit power Rayleigh process, whose dynamics are obtained by setting x_t equal to a constant of unity and $\mathcal{A} = 0$ in the above system.

(The original proof of this result appears as Proposition 2.1 in Field and Tough 2003b, and we refer the reader to Chapter 8 herein for a detailed mathematical derivation.) Thus, the non-linear SDE for Ψ_t is derived theoretically from first principles beginning with the random walk model for the scattered electric field under the assumption of a uniform phase distribution.[39] An immediate consequence of this dynamical equation is the 'noise-free skeleton', obtained by setting the volatility coefficients of the fluctuating Wiener terms equal to zero. Accordingly, the randomness of the process is eliminated and the residual dynamics are deterministic and differentiable. Physically, this corresponds to an evolution conditioned on the current state of the system and then averaged over an ensemble.[40] The concept of the residual noise-free part is explored further below.

The constants $\mathcal{A}, \mathcal{B}$ in (13.5), (13.6) have the physical dimension of frequency, so that their reciprocals represent correlation timescales for the RCS modulation and Rayleigh scattered components, respectively. The constant $\mathcal{B}$ is electromagnetic in origin with a value $\mathcal{B} \sim c|\mathbf{k}|$ where $\mathbf{k}$ is the wave vector of the carrier. In radar situations, the illuminating radiation is such that $\mathcal{A} \ll \mathcal{B}$, with the value of $\mathcal{A}$ being determined as an intrinsic property of the statistics of the scattering

---

[38] We can express $\Psi_t = I_t + jQ_t$ ($j = \sqrt{-1}$), the familiar sum of its 'in-phase' and 'quadrature-phase' components.

[39] The assumption of a uniform phase distribution can be relaxed and a corresponding detailed dynamical description in terms of SDEs has been given in Field and Tough (2005).

[40] In other words, for an Ito process q_t with SDE $\mathrm{d}q_t = b_t\mathrm{d}t + \sigma_t\mathrm{d}W_t$, the ensemble average evolution is determined by $\mathbf{E}[\mathrm{d}q_t] = b_t\mathrm{d}t$.

surface, independent of the electromagnetic wave. Accordingly in radar the correlation time for the RCS is much longer than that of the Rayleigh speckle. (The reader should compare also the discussion of amplitude and frequency modulation given at the end of this section.)

This set of coupled stochastic dynamical equations is manifestly *non-linear* by virtue of the reciprocal term in x_t appearing in the amplitude equation, and only reduces to linear dynamics in the special case that $\mathcal{A}$ vanishes, i.e. the scattering cross-section is constant (Rayleigh scattering).

13.3 Radar parameters

In the present context, it is worth remarking on some of the key salient features of the SDE theory, in relation to the sensitivity analysis of sea clutter to certain radar parameters. Most notably, this kind of description is illuminating in respect of the following issues.

13.3.1 *Superposition*

In light of the SDE theory, we may argue that the SDE of sea clutter is independent of the amplitude profile of a transmitted pulse, provided the transmit energy is maintained constant. This property, which is derived explicitly in Field (2005), is related to the fact that the form factors (i.e. the amplitude weightings) in (13.4) may be taken as unity for an asymptotically large population (cf. also Jakeman and Tough 1988 where the emergent statistical properties are independent of the choice of form factors).

For a radar pulse of constant amplitude, suppose that the two halves of the pulse have transmit frequencies ω_1 and ω_2. Then we may consider the correlation between the SDE of sea clutter for the two portions as frequency ω_1 increases relative to ω_2. The transmit frequencies are proportional to the Rayleigh constant $\mathcal{B}$ appearing in (13.6) ($\mathcal{B} \sim c|\mathbf{k}|$, c is the speed of light, $\mathbf{k}$ is the carrier wave vector), and the relationship between the two SDEs, for the two different transmit frequencies, is through (13.6): the two terms involving the constant $\mathcal{A}$ are the same for both SDEs; on the other hand, the terms involving the Rayleigh constant have different $\mathcal{B}$ values corresponding to the two transmit frequencies. Nevertheless, on physical grounds, the two complex Wiener processes ξ_t for each transmit frequency should be considered perfectly correlated. The reason for this correlation is that the physical origin of the component phase fluctuations $\phi^{(j)}$ is (microscopic) Doppler – the Doppler frequency ratio ω_1/ω_2 is a function of the radial velocity of the j-th member of the population, so the micro-Doppler phase shift scales with the transmit frequency; the ξ_t process is the same for any transmit frequency (assuming these are transmitted simultaneously) as this depends only on the behaviour of the component scatterer.

In a similar fashion, consider the simultaneous transmission of two pulses of constant amplitude, with two different frequencies as above, and the resulting SDE of sea clutter received by a common antenna. Since Maxwell's equations of electromagnetism are linear, the resulting Ψ is a linear superposition $\Psi =$

$k_1 \Psi^{(1)} + k_2 \Psi^{(2)}$ where k_1, k_2 are the relative intensities of the two transmit waveforms, normalized so that $k_1 + k_2 = 1$, and $\Psi^{(i)}$ are the constituent complex amplitude processes, both satisfying the SDE (13.6), with different Rayleigh constants $\mathcal{B}$ corresponding to the two transmit frequencies. Since the beams are simultaneous, the ξ processes are perfectly correlated, with the remaining parts of (13.6) involving the constant $\mathcal{A}$ being the same for both transmit frequencies. Thus, the non-linear dynamics do not infringe the principle of superposition inherent in Maxwell's equations.[41]

13.3.2 Sea state and polarization

Next, consider the two different like-polarizations 'HH' and 'VV'. The SDE theory encapsulates the spikes in the RCS of sea clutter due to 'HH' versus the noise-like character due to 'VV' as follows. The cross-section SDE (13.5) contains the parameter α, where $\alpha = \nu/\lambda$ and λ, ν are the birth and immigration parameters of the underlying (BDI) population model, respectively. Now $\mathrm{Var}[x] = \mathbf{E}[x] = \alpha$, so the absolute magnitude of fluctuations in the RCS, that give rise to the K-distribution for the intensity (as opposed to the Rayleigh 'noise-like' distribution), become more appreciable as α increases. However, the appropriate theoretical measure of 'spikiness' is the *relative* variance R given by

$$R = \frac{\mathrm{Var}[x]}{(\mathbf{E}[x])^2} \qquad (13.7)$$

which is the physical parameter of interest since it is dimensionless and invariant under re-scaling of the RCS. In the case of K-scattering that we consider here, R is equal to $1/\alpha$, and therefore the horizontal 'HH' like-polarization has small α, with larger α for vertical like-polarization 'VV'. The SDE theory explains that if the ratio α of the immigration to birth rates is small, then the sea clutter possesses spikes. It is therefore a natural mathematical, as opposed to a detailed phenomenological, way of encoding this physical property of the sea surface.[42] Correspondingly, there are two different K-distributions for the intensity, indexed by different values of the shape parameter $\nu = \alpha - 1$, for the respective polarizations.

The situation as regards the extent of the temporal fluctuations in the RCS for low/moderate/high values of the shape parameter is well illustrated in Fig. 13.2. The figure demonstrates the extreme departures from the mean value for large R, which represents, in physical terms, sea spikes or glints in the scattering surface. The dynamical nature of these extreme departures can be understood further, to some extent, by calculating the SDE of the normalized RCS $x' = x/\alpha$, which is seen to be

---

[41] It is necessary to assume here that the scattering populations $N^{(1)}$ and $N^{(2)}$ pertaining to the different transmit frequencies are equal.

[42] However, the SDE theory does not explain why for 'HH' polarization one should expect the population to behave this way, the phenomenological reasons for which we do not describe here.

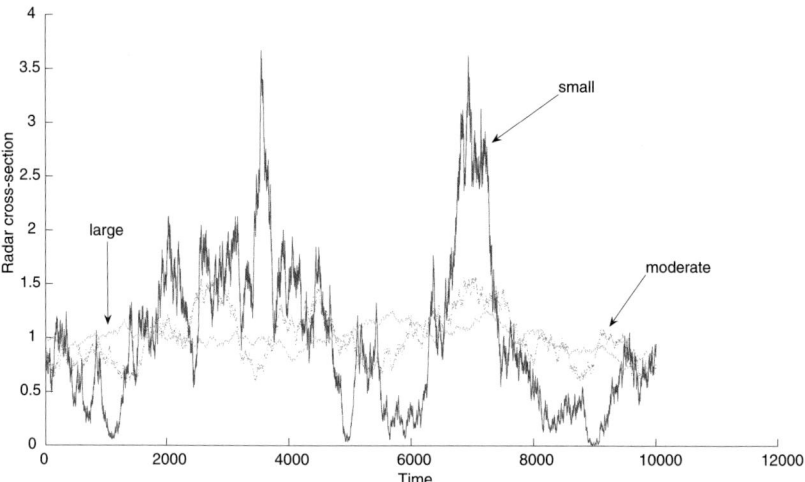

FIG. 13.2. Normalized RCS time-series for low/moderate/high values of the shape parameter; simulated data with $\mathcal{A}\delta t = 0.001$, $\alpha = 1, 10, 100$.

$$\mathrm{d}x'_t = \mathcal{A}(1 - x'_t)\mathrm{d}t + \left(\frac{2\mathcal{A}x'_t}{\alpha}\right)^{\frac{1}{2}} \mathrm{d}W_t^{(x)}. \tag{13.8}$$

A natural question that arises for the domain of low α (high sea state) is whether, following a burst, the RCS returns to equilibrium and undergoes a period of relative quiescence before a successive burst, or whether the high relative variance is achieved through perpetual bursts of varying magnitude, with the maximum timescale for their separation (pertaining to the largest deviations) being given by the reciprocal of the constant $\mathcal{A}$. Inspection of (13.8) reveals the latter situation to be the case, as setting $x' \approx 1$ therein removes the drift term but nevertheless produces a highly volatile behaviour for small values of α. Observe also, in this regard, the essential difference between the roles of $\mathcal{A}$ and α, which determine the correlation timescale and magnitude of excursions respectively – thus, for instance, the single point distribution of the RCS is unaffected by $\mathcal{A}$, whereas it is α that parameterizes this distribution according to the discussion of the shape parameter above. These features are indeed evident from the simulated data shown in Fig. 13.2 (for $\alpha = 1$). Observe also, embedded within the largest peak to peak timescale $O(\mathcal{A}^{-1})$, the presence of sub-scale peaks separated by times of smaller order, including scales less than the characteristic Rayleigh timescale $O(\mathcal{B}^{-1})$.

Now, in contrast, as the sea state settles down to a low value, the (normalized) RCS has small fluctuations away from its mean (unity), so that there is no significant modulation of the Rayleigh scattering time series – in other words, the scattering is approximately of constant local power.

As the sea state diminishes, correspondingly in terms of the SDE dynamics the parameter $\alpha \to \infty$ and the relative variance in the RCS tends to zero. Thus in Fig. 13.2 the non-linear term becomes less pronounced. Accordingly, as we have seen, so does the degree of non-linearity as measured by the z-parameter as developed in Section 13.1, which confirms the experimental findings reported in Haykin et al. (2002).

We close this chapter with some general discussion concerning the main aspects of the non-linear dynamics of sea clutter. We have described a detailed analysis of radar sea clutter data, whose primary purpose is to address the presence of non-linearity, from real experimental data. A natural quantifier for this non-linearity is the z-parameter or Mann–Whitney rank-sum statistic, which has been successfully applied in the context of a hybrid AM/FM model for sea clutter. The SDE dynamical model of radar sea clutter has also been verified previously to a remarkable degree of accuracy, in terms of real experimental data (*orig.* section 4(*b*) in Field and Tough 2003*a*). Moreover, an independent theoretical account for such a model was provided in Field and Tough (2003*b*), and has served as the basis for other significant developments (Field and Tough 2005; Field 2005). As we have seen in Section 13.2, this stochastic dynamic behaviour is inherently non-linear, due to the broader timescale fluctuations in the RCS. The extent of non-linearity arises naturally in the SDE description through the relative variance or shape-parameter, which encodes the sea state. Thus, from an SDE dynamical perspective, the non-linear character of radar sea clutter is firmly established, both theoretically and experimentally.

Calculation of the z-parameter is from real data *containing* noise, the latter being akin to the stochastic fluctuating terms present in (13.5), (13.6). However, z has the stochastic element removed, i.e., it is not a random variable. Accordingly, some ensemble averaging takes place in the calculation of z, and for this purpose the statistical properties of ergodicity and stationarity are assumed, legitimate over realistic short timescales. In terms of the parameter $\mathcal{A}$ of (13.5), such timescales are short enough that the assumption of constant $\mathcal{A}$ is valid. Nevertheless, they should be long enough (of the order of $\mathcal{A}^{-1}$) for the fluctuations in the RCS (or equivalently, as we elucidate below, the frequency modulation effect) to be appreciable so that non-linearity can indeed be detected.

From an engineering physics perspective, the dynamics of sea clutter are perhaps more naturally viewed in terms of amplitude (AM) and frequency modulation (FM). Studies have indicated that the degree of non-linearity is governed by the extent of FM which, in turn, is more noticeable for higher sea states (i.e. the shape parameter ν is large). To relate this further to the SDE description of Section 13.2, it is convenient to view the resultant amplitude process Ψ_t in the product representation $\Psi = x^{1/2}\gamma$, in which x is the RCS and γ is a unit power Rayleigh process. Then, the AM consists of the fluctuations of γ_t (Rayleigh 'speckle') which is 'frequency' modulated by the RCS process x_t over a much broader timescale. The FM/AM contributions therefore have characteristic frequencies determined by $\mathcal{A}$, $\mathcal{B}$ respectively. With zero FM, i.e. x_t constant, the dynamics of the resultant amplitude are rendered linear, according to Proposition 13.1.

Observe that, whereas the FM/AM characteristics of the received envelope do not map to unique stochastic dynamics, conversely the SDE description allows for explicit extraction of both the FM/AM constituents (see Field 2005), and therefore the SDE description is more fundamental than the spectral one. Indeed, given the SDE dynamics, we are able to extract all higher order statistical information through the propagators obtained as solutions of the associated Fokker–Planck equations (Risken 1989; Field and Tough 2003b). In this way, the SDE description of sea clutter should be viewed as the most complete dynamical description, which preserves the inherent randomness in the physical processes involved.

We have seen that independent lines of enquiry, from theoretical and experimental perspectives, lead to the common conclusion that radar sea clutter is non-linear over appreciable timescales such that the temporal variation in the RCS is significant. The degree of non-linearity is determined by the sea state, which is represented by a certain 'shape parameter' ν that features in the SDE for the RCS.[43] Consistently, the non-linearity is also determined by the extent of frequency modulation which, in terms of real experimental data, has been quantified in terms of a certain z-parameter representing the Mann–Whitney rank-sum statistic.

From a theoretical point of view the deterministic part of the stochastic dynamics (13.5), (13.6) is non-linear, and is augmented with the addition of fluctuating Wiener terms in the description of real experimental data, which is inherently noisy. (The reader should compare the discussion of chaos surrounding figure 2 in Sugihara 1994, and also Stone 1992, Sugihara and May 1990.) We recommend further studies be made on the noise-free skeleton of the coupled system (13.5), (13.6), which is manifestly non-linear, to establish the existence or otherwise of an underlying deterministic chaotic behaviour. If chaos is present, then this system of non-linearly coupled SDEs is an instance of 'stochastic chaos'. We remark in this respect that the presence of the Wiener fluctuating terms in the system has the effect of stabilizing the system, so that any chaotic behaviour may no longer be observable experimentally. These issues will be pursued elsewhere in the journal literature.

It is worth emphasizing again that the SDE theory of sea clutter is experimentally valid, in its own terms (*orig.* section 4(b) in Field and Tough 2003a), and has also succeeded in practical applications, such as radar anomaly detection, to a remarkable degree of accuracy. The theory also provides a way of generating synthetic data, over which we have direct control, in terms of its dependence on the sea state. Thus, in principle, we could measure the z-parameter for a data

---

[43]More precisely, $\nu = \alpha - 1$ where α is the parameter in the SDE for the RCS, and $\alpha \geq 0$ arises from the parameters in the scattering population model as described in Chapter 7.

set simulated using SDEs, for which the shape parameter is known, and thereby develop the precise relationship between the z-parameter and shape parameter quantifiers of non-linearity. It may, indeed, also be possible to relate the two parameters in purely theoretical terms. We can also generate data with and without noise, which forms the basis for further experiment. We suggest that these two lines of enquiry could form the basis of future developments in the investigation of the non-linear properties of radar scattering dynamics.

14

OBSERVABILITY OF SCATTERING CROSS-SECTION

A study of some synthetically generated data provides a clear illustration of the consequences of Theorem 10.6 and Proposition 10.7 in a variety of experimental contexts. The theoretical results are conveniently substantiated using synthetic data, as such enables direct verification of the accuracy of the state estimation from the observations, the precise value of the 'hidden' state being known. One can then compare the values of the state inferred from the observations alone, with the exact values of the underlying hidden state recorded in the simulation.

14.1 Simulated data

The cross-section was chosen to satisfy the stochastic differential equation (SDE) (10.5) with $b(t,x) = \alpha - x$, $\Sigma(t,x) = x$ so that x_t is a gamma variate and the process thus generated is appropriate to the types of scattering data found in radar applications (cf. Field and Tough 2003a). Resultant amplitude data was simulated via integration of (10.7), which is achieved most effectively by a separate numerical integration of the (independent) component SDEs (10.3) and (10.5). We emphasize however, that the same types of numerical results as demonstrated below should hold for an arbitrary population, as shown theoretically in Theorem 10.6. For the purposes of the simulation, α was chosen to be large, to avoid numerical difficulties that can arise due to the singular behaviour in the phase fluctuations at zeros of the intensity, implied by (10.15). The results of the simulation are provided in Fig. 14.1, which shows time series for the observed intensity, the exact cross-section generated in the simulation (i.e. the unknown state one is trying to estimate), and the values of this state inferred from the observations of the scattered amplitude alone. The estimate of the state follows from Theorem 10.6 which, for discretely sampled data, implies

$$z_i \delta \theta_i^2 \propto x_i n_i^2 \qquad (14.1)$$

where i is a discrete time index and $\{n_i\}$ are an independent collection of $\mathcal{N}(0,1)$ distributed random variables. Applying a smoothing average $\langle \cdot \rangle_\Delta$ to the left-hand side (the 'observations') of (14.1) with window $\Delta = [t_0 - \Delta, t_0 + \Delta]$ yields an approximation to x_{t_0}, with an error that tends to zero as the number of pulses inside Δ tends to infinity and $\Delta \to 0$ (see discussion of χ^2 statistic following Theorem 10.6).[44]

---

[44] We assume here that the sample paths of x_t are continuous a.s., which is a consequence of (10.5).

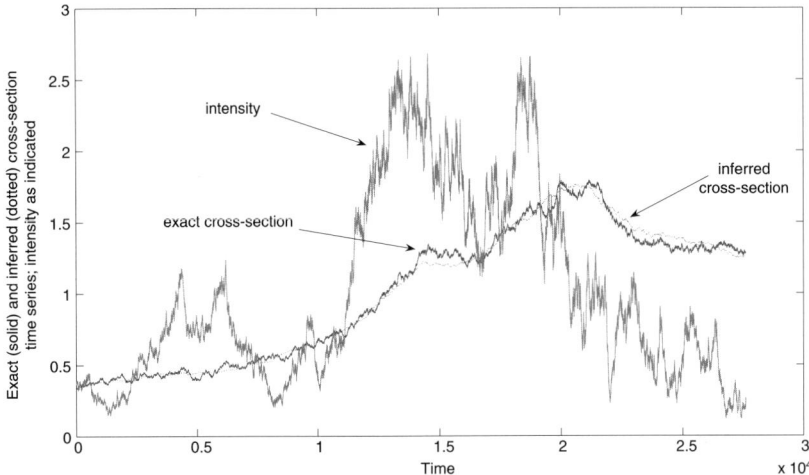

FIG. 14.1. Estimation of the scattering cross-section/population after relaxation time T through the effect of phase decoherence. (For parameter values $\alpha = 10$, $\Delta = 1200$ pulses, $\mathcal{A} = 10^{-5}$, $\mathcal{B} = 10^{-4}$ the figure shows a statistical correlation between the exact and inferred cross-section time series of 0.9954.)

Remarks. *On finite pulse rate.* In the case that the pulse rate is bounded, as in a real experiment, one should consider the optimization of the smoothing window – the window must be neither so large that the structure of the temporal variation in the cross-section is lost, nor so small that an average over the normal random seeds is no longer effected. Thus a compromise optimal value should exist, and can indeed be derived (at least for K-scattering). We refer the interested reader to Fayard and Field (2008) for a detailed account.

On theother hand, smoothing the z_t time series, for any choice of parameter Δ, does not yield the desired close correlation with x_t; indeed any such attempt to 'decorrelate the speckle pattern' merely produces an intensity profile with the same general shape as the original z_t, with oscillations on a timescale approximately equal to Δ.

14.2 Experimental applications

The application of related earlier ideas in SDE theory to optical propagation and radar scattering has been reported in §4 of Field and Tough (2003*a*). The current results should be of additional value in these types of experiments, through the ability to estimate the underlying scattering cross-section for general scattering populations, in real time. When applied to experimental data, the results herein provide means for studying the behaviour of random media based on statistical analysis of the inferred cross-section alone, which has hitherto been regarded as the 'hidden' state of the system. The electromagnetic scattering process should then appropriately be viewed as a secondary exogenous device, whose purpose is merely to extract the real time behaviour of the underlying scattering cross-section, where the latter is the object of primary interest.

Our results also suggest the application to the physics of magnetic resonance (MR) imaging and spectroscopy and experimental NMR (Field and Bain 2008). The random medium (e.g. brain tissue) lies inside a background magnetic field B_0, with which the constituent (proton) spin vectors are aligned, in their minimum energy configuration. An applied RF pulse causes resonant absorption to occur, so that the spins re-align, typically at a 'pulse flip' angle of 90° to B_0. Radiation of this absorbed energy gives rise to the received MR signal (the 'free induction decay' or FID), which is detected through the generation of electromotive force in a coil apparatus, due to the time varying local magnetic field.[45] The MR signal has the usual in-phase (I) and quadrature-phase (Q) components familiar from radio theory, and thus corresponds to the amplitude process $\psi = I + iQ$ for each point in space. For a perfectly homogeneous (total) magnetic field throughout the medium, each spin vector precesses at the Larmor frequency ω_0 about the longitudinal axis, where ω_0 is given by the Larmor equation $\omega_0 = \gamma B_0$ and γ is the (local) gyromagnetic ratio (e.g. Ernst et al. 1987). (In the radar scattering situation described above, ω_0 corresponds to the Doppler frequency, arising from bulk wave motion in the scattering surface.) However, the local *in*homogeneities in the net magnetic field, due to the local magnetic properties of the medium, give rise to a process known as 'spin-spin' or '$T2$ relaxation' constituting the (random) exchange of energy between neighbouring spins. These local perturbations in the total magnetic field can reasonably be considered as independent for each component spin, so that the dynamics of each spin vector can be modelled as a phase diffusion process with (transverse) resultant as in (6.16) and phase initializations $\{\Delta^{(j)}\}$ equal. After sufficient relaxation time $t \geq T$ has elapsed, phase decoherence occurs. In principle, this enables the spin population to be tracked according to Theorem 10.6, and thus real-time MR images to be generated. This application is explored in greater depth, and from a less detailed mathematical perspective, in Field (2006). In the context of experimental NMR these theoretical ideas have been tested and verified, as reported in Field and Bain (2008).

---

[45] This effect is the result of Faraday's law, i.e. Maxwell's curl equation for a time dependent magnetic field $\nabla \times \mathbf{E} = -\partial \mathbf{B}/\partial t$ integrated around a loop.

APPENDIX A

STABILITY AND INFINITE DIVISIBILITY

First, let us remark on the basic properties of sums of independent random variables. Suppose $\{X_i\}$ is a collection of independent continuous random variables with density functions p_{X_i}. Let Z be their sum, $Z = \sum X_i$. Then the distribution of Z is given by the n-fold convolution, namely

$$p_Z = p_{X_1} * p_{X_2} * \cdots * p_{X_n} \tag{A.1}$$

in which the convolution of two functions f, g is defined as usual according to $f * g(x) = \int f(u) g(x - u) \mathrm{d}u$. Correspondingly, in terms of the characteristic function, defined by $\Phi_X := \mathbf{E}\left[e^{i\omega X}\right]$, we have the n-fold product

$$\Phi_Z(\omega) = \prod_1^n \Phi_{X_i}(\omega). \tag{A.2}$$

(The same formulae, of course, do not hold if X_i are statistically dependent.)

Consistently with the nomenclature, the property of stability concerns *superposition* while infinite divisibility concerns *decomposition* of random variables (distributions), and the component random variables that arise in the definitions of each property are considered independent and identically distributed. Strictly speaking, the two properties should correctly be considered as pertaining to distributions (not random variables).

A.0.0.1 Infinite divisibility A distribution F is said to be *infinitely divisible* if, for each positive integer n, it arises as the n-fold convolution of a (distinct) distribution $F^{(n)}$ with itself, i.e.

$$F = \overbrace{F^{(n)} * \cdots * F^{(n)}}^{n \text{ terms}}. \tag{A.3}$$

A.0.0.2 Stability The distribution F is said to be *stable* if, given $\{X_i\}$ independent and $\{X_i, X\}$ identically distributed according to F, then for all choices of constants a, b there exist constants c, d such that

$$aX_1 + bX_2 \doteq cX + d \tag{A.4}$$

with equality $\doteq$ in distribution.

By expressing these conditions for infinite divisibility and stability separately in terms of characteristic functions, as explained above, one can deduce that stability implies infinite divisibility. The converse does not hold, however. Some pertinent cases are the following:

Gaussian. The Gaussian distribution is stable. This follows from the distribution properties:

$$X \sim \mathcal{N}(\mu, \sigma^2),$$

$$p_X(x) = \frac{1}{\sqrt{2\pi\sigma^2}} \exp(-(x-\mu)^2/2\sigma^2),$$

$$\Phi_X(\omega) = \exp\left[i\mu\omega - \sigma^2\omega^2/2\right].$$

Gamma. The gamma distribution, with basic properties

$$X \sim G(\alpha, \beta),$$

$$p_X(x) = \frac{x^{\alpha-1}e^{-x/\beta}}{\Gamma(\alpha)\beta^\alpha}; \quad x \geq 0; \quad \alpha, \beta > 0,$$

$$\Gamma(z+1) = z\Gamma(z),$$

$$\Phi_X(\omega) = (1 - i\omega\beta)^{-\alpha}$$

is infinitely divisible. However, consistently with the above general observations, it is not stable (as apparent from the characteristic functional form).

Cauchy

$$X \sim G(\alpha, \beta),$$

$$p_X(x) = \frac{1}{\pi} \frac{a}{a^2 + (x-b)^2},$$

$$\Phi_X(\omega) = \exp(ib\omega - a|\omega|).$$

This distribution (also referred to as the 'Lorentzian' distribution, of importance in resonance theory) is stable (and therefore, like the Gaussian, also infinitely divisible). Observe also that (for $b = 0$) the distribution of the sample mean $\mu_N = \sum_{i=1}^{N} X_i/N$ where X_i are drawn independently from a Cauchy distribution, has the *same distribution* as X_i, so the ordinary central limit theorem drastically fails.

On the basis of the same type of reasoning the K-distribution encountered in the description of K-scattering is also infinitely divisible (but not stable). Of mathematical interest, distributions for discrete multi-valued random variables are neither infinitely divisible nor stable (as can be seen by considering the possible sample values that occur under linear combinations as compared to those drawn from the original distribution).

APPENDIX B

ITO VERSUS STRATONOVICH STOCHASTIC INTEGRALS

Let us compare the decompositions of the differential of a stochastic process X_t in Ito versus Stratonovich terms:

$$dX_t = \begin{cases} b^{(I)}dt + \sigma dW_t \\ b^{(S)}dt + \sigma \circ dW_t \end{cases} \quad (B.1)$$

in which $b^{(I)}$, $b^{(S)}$ denote the drifts in the Ito and Stratonovich interpretations respectively and '$\circ$' is a shorthand that indicates the Stratonovich prescription for taking the stochastic integral, i.e. that the volatility is evaluated at the *midpoint* of each subinterval. Now we can study the Stratonovich stochastic integral term above and translate it into the Ito interpretation:

$$\sigma \circ dW_t = \left(\sigma + \frac{1}{2}(\sigma \partial_x \sigma)dW_t\right) dW_t + o(dt^{1/2}) \quad (B.2)$$

in which the origin of the factor $\frac{1}{2}$ on the right-hand side is that the midpoint of the subinterval is being applied, in the sense explained above.[46] Thus, neglecting terms of $o(dt^{1/2})$ we deduce that the Stratonovich and Ito stochastic integrands are related by

$$\sigma \circ dW = \sigma dW + \frac{1}{2}\sigma \partial_x \sigma dt \quad (B.3)$$

which follows from the relation $dW_t^2 = dt$. Hence we find that the two notions of drift are related by

$$b^{(I)} = b^{(S)} + \frac{1}{2}\sigma \partial_x \sigma. \quad (B.4)$$

It is important to appreciate that the stochastic process X_t can be consistently represented *either* way, provided one is clear which prescription for the stochastic integral is being applied. Translation between the two representations occurs with respect to the drift (only), and is according to (B.4).

---

[46] A corresponding construction holds if the volatility is evaluated at $t_i + \alpha \delta t$ in the interval $[t_i, t_i + \delta t]$ thus generating a 1-parameter family of stochastic integrals – the cases of Ito and Stratonovich then correspond to α equal to zero and $\frac{1}{2}$ respectively.

It is of mathematical interest that the Stratonovich interpretation of the stochastic integral leads to a calculus for which the ordinary rules of differentiation apply. More precisely, for example, the 'chain rule' is valid so that, in the Stratonovich interpretation, we have the classical differential formula

$$dF_t = \frac{\partial f}{\partial X} \circ dX_t. \tag{B.5}$$

This result follows from the same type of argument as in (B.2) applied to $f' \circ (\sigma dW)$, integrating the f' factor over a semi-interval, which yields the second-order derivative term $\frac{1}{2}f''\sigma^2 dt$ that is characteristic of the Ito calculus.

We introduce the *martingale* property for a stochastic process M_t

$$\mathbf{E}[M_t|\mathcal{F}_s] = M_s, \quad s \leq t. \tag{B.6}$$

Observe that the expectation of an Ito stochastic integral vanishes, i.e.

$$\mathbf{E}\left[\int_a^b \phi_s dW_s\right] \equiv 0. \tag{B.7}$$

Thus, Ito stochastic integrals are martingales, unlike their Stratonovich counterparts.

Finally, comparing with the development of stochastic differential geometry in Chapter 3, first recall the relation between the Ito and Kolmogorov drifts on $(\mathcal{M}, \sigma^{ij})$

$$b^i = \beta^i + \frac{1}{2}\Gamma^i \tag{B.8}$$

where $\Gamma^i = \sigma^{jk}\Gamma^i_{jk}$, the trace of the Christoffel symbol. Then in one-dimension

$$\Gamma^1 = -\sigma \partial_x \sigma. \tag{B.9}$$

Accordingly, we should compare (B.4) with the one-dimensional instance of (B.8), namely

$$\beta = b + \frac{1}{2}\sigma \partial_x \sigma. \tag{B.10}$$

To see the consistency of these two points of view, observe that the Stratonovich drift can be obtained (at a point of $\mathcal{M}$) in a normal coordinated system (with respect to σ_{ij}) and that the Stratonovich drift transforms homogeneously under coordinate transformations $x^i \mapsto x^{\hat{i}}$.

The interested mathematical reader should also compare Wong and Zakai (1965) for a related discussion in the context of the relationship between ordinary and stochastic differential equations (SDEs). We can consider an approximate solution to the SDE (B.1) generated by a sequence $x_t^{(n)}$ of solutions to the ordinary differential equation $dx^{(n)}/dt = b + \sigma\Gamma^{(n)}$ in which $\Gamma_t^{(n)}$ is a sequence of processes whose integral $W_t^{(n)} = \int^t \Gamma_s^{(n)} ds$ converges to the Wiener process as $n \to \infty$. Then the solution $x^{(n)}$ converges to the *Stratonovich* solution of the SDE.

Remarks. *On numerical integration.* Sometimes it may be appropriate from a numerical stability point of view to use a numerical integration (discrete time step) scheme based on using the midpoint of each subinterval, the results of which therefore converge to the solution of the *Stratonovich* SDE. When confronted with an Ito SDE (B.1) with a volatility that is state dependent (and possibly non-linear) one can, according to the above results, proceed to an accurate numerical solution via a *midpoint* method applied to the following (Stratonovich) SDE, which contains a modified drift according to the transformation (B.4),

$$dX_t = \left(b^{(I)} - \frac{1}{2}\frac{\partial \sigma}{\partial x}\right) dt + \sigma \circ dW_t. \qquad (B.11)$$

The solution thus generated is the desired approximation to the original Ito SDE.

Throughout the monograph the *Ito interpretation* is consistently adopted.

APPENDIX C

FILTRATIONS, CONDITIONAL PROBABILITY, AND MARKOV PROPERTY

We let $\mathbf{E}[\cdot|\mathcal{G}]$ denote the expectation functional conditional on information represented by (the σ-algebra) $\mathcal{G}$. Then if $\mathcal{G} \subset \mathcal{H}$, i.e. the *information* represented by $\mathcal{G}$ is contained in that of $\mathcal{H}$, then[47] we have the *iterated expectation* rule

$$\mathbf{E}\left[\mathbf{E}[\cdot|\mathcal{F}_t]|\mathcal{F}_s\right] \equiv \mathbf{E}[\cdot|\mathcal{F}_s]. \quad (C.1)$$

In terms of *time* conditioned iterated expectations for a stochastic process X_u we introduce the abstract notion of a 'filtration' $\mathcal{F}_s$ which (is a σ-algebra that) represents the information contained in the history of the process up to and including time s. Now, if we set $\mathcal{G}$, $\mathcal{H}$ to correspond to $\mathcal{F}_s$, $\mathcal{F}_t$ respectively, for $s \leq t$, then we obtain the *tower law*

$$\mathbf{E}_s \circ \mathbf{E}_t \equiv \mathbf{E}_s, \quad s \leq t \quad (C.2)$$

in which $\mathbf{E}_s[\cdot] \equiv \mathbf{E}[\cdot|\mathcal{F}_s]$.

If $\mathbf{E}_s$ is defined as above, so that it is conditional on the *history* of the process, then the tower law holds for a general process. Observe, on the other hand, if we define $\check{\mathbf{E}}_s$ as the expectation conditioned only *at* the instant s, then we require the *Markov property* for the corresponding tower law to hold for $\check{\mathbf{E}}$, i.e. that

$$\check{\mathbf{E}}_s X_u \equiv \mathbf{E}[X_u|\mathcal{F}_s]. \quad (C.3)$$

---

[47]This follows essentially from the probability rule $\mathbf{P}[A|B]\mathbf{P}[B] \equiv \mathbf{P}[A \cap B]$.

APPENDIX D

GIRSANOV'S THEOREM

Consider a process X_t that satisfies the stochastic differential equation (SDE)

$$\mathrm{d}X_t = b_t \mathrm{d}t + \mathrm{d}W_t, \qquad (\text{D.1})$$

where W_t is a Wiener process, with respect to the probability measure **P**.

Consider now the process $Y_t := M_t X_t$ where M_t is the *exponential change of measure* defined as

$$M_t := \exp\left(-\int_0^t b_s \mathrm{d}W_s - \frac{1}{2}\int_0^t b_s^2 \mathrm{d}s\right). \qquad (\text{D.2})$$

We deduce, via Ito's formula, the stochastic differential of M_t

$$\mathrm{d}M_t = -b_t M_t \mathrm{d}W_t \qquad (\text{D.3})$$

and thus M_t is itself a martingale with respect to the measure **P** (it has vanishing drift). It follows that the stochastic differential of Y_t is given, according to the Ito product rule, by

$$\mathrm{d}Y_t = M_t(1 - b_t X_t)\mathrm{d}W_t \qquad (\text{D.4})$$

and hence Y_t is also a **P**-martingale.

This line of argument leads to Girsanov's theorem, which states that[48] if we make the *Girsanov transformation of measure* from **P** to **Q** on the space of paths according to

$$\mathrm{d}\mathbf{Q} = M_T \mathrm{d}\mathbf{P}, \qquad (\text{D.5})$$

i.e. with Radon–Nikodym derivative $\mathrm{d}\mathbf{Q}/\mathrm{d}\mathbf{P} = M_T$, then X_t is a Wiener process with respect to the transformed measure **Q** defined on $\mathcal{F}_T$.

Intuitively, the theorem says that if we re-assign the probabilities of paths according to **Q**, then with respect to this (ensemble average) measure the drift of the process (the tendency for X to increase or decrease, on average) is eliminated. This enables us to consider a single path of a stochastic process as an instance of its drift-free counterpart. In other words, from the point of view of instantaneous observable features of the process, the drift has no significance. It is only when

---

[48] subject to existence of the change of measure, which is ensured by the *Novikov* condition, $\mathbf{E}\left[\exp\left(\frac{1}{2}\int_0^T b_s^2 \mathrm{d}s\right)\right] < \infty$.

ensemble averages are applied that the drift concept enters into the description and, in real time physics applications such as we have considered, it may be that no such ensemble average is practicable or indeed available.

In contrast to the drift, the volatility is invariant under changes of measure on path space. This is the mathematical basis of the stochastic volatility analysis applied to scattering, and implies that no ensemble average is required to determine the volatility from a single realization (path) of a stochastic process.

Remarks. *On coin tossing.* As a useful simple analogy, we could consider tossing a weighted coin – generate a (discrete-time/valued) process by starting at the origin and moving up/down according as heads/tails is obtained. Suppose now we observe a *single* path of this process, and that it tends to move upwards much more than downwards. Then consider the question of whether, given the observation of the single path, the coin is biased toward heads (positive drift) or is a fair coin (no drift). The answer to this question is, of course, *indeterminate* – the coin could either be fair and the observed path is an unlikely one, or it could be a 'typical' path for a head-weighted coin. Indeed, one can see this intuition reflected in the explicit expression for the change of measure (D.5) which suppresses probabilities of paths with large drift.

D.0.1 *Relation to mathematical finance*

We remark briefly that exactly the same principle applies to financial data, which is observed and hedged in real time.[49] Accordingly the Black–Scholes option price is drift independent, obtained in the 'risk-neutral' measure. The use of this measure, as opposed to the original 'real world' measure, amounts to the distinction between arbitrage versus expectation pricing. It is the former strategy for which the Black–Scholes theory provides an explicit option price, and which enables an individual in a market to eliminate risk in a random environment, by hedging the observed stock based on their instantaneous values/fluctuations.

We refer the interested mathematical reader to Oksendal (1998) for a more thorough and rigourous account of Girsanov's theorem.

---

[49]These remarks may be omitted without interrupting the main flow of the development. They are mentioned for the purpose of highlighting the intimate connections that exist between mathematical physics and financial mathematics, and hopefully therefore to stimulate interaction between the two scientific communities.

APPENDIX E

PARTITION FUNCTION SOLUTION TO BDI MODEL

The 'partition function' for the birth–death–immigration (BDI) process satisfies the following differential equation (Bartlett 1966):

$$\frac{\partial \Pi_t(z)}{\partial t} = (z-1)(\lambda z - \mu)\frac{\partial \Pi_t(z)}{\partial z} - \nu(1-z)\Pi_t(z) \quad \text{(E.1)}$$

in which the partition function is defined by $\Pi_t(z) = \langle z^{N_t} \rangle$.[50] This equation has the finite-time closed form solution

$$\Pi_t(z) = \frac{(\lambda - \mu)^{\nu/\lambda}[\mu(T-1) - (\mu T - \lambda)z]^n}{[\lambda T - \mu - \lambda(T-1)z]^{n+\nu/\lambda}} \quad \text{(E.2)}$$

in which we abbreviate $T = \exp((\lambda - \mu)t)$ and n is the specified number in the population initially. The condition for the existence of an asymptotic equilibrium solution (distribution) is $\mu > \lambda$ which ensures that the function $T(t)$ decays to zero as $t \to \infty$. In this limiting case, recall that the negative binomial variate N has probability mass function

$$p_N = \binom{\alpha + N - 1}{N} p^\alpha q^N \quad \text{(E.3)}$$

for integer $N \geq 0$ with partition function

$$\Pi_\infty(z) = \left(\frac{p}{1-qz}\right)^\alpha \quad \text{(E.4)}$$

to be compared with the special case of the Poisson variate for which

$$\Pi_\infty(z) = \exp[-\lambda(1-z)]. \quad \text{(E.5)}$$

Thus in terms of the BDI parameters, we identify $p = 1 - \lambda/\mu$, $q = \lambda/\mu$ ($p+q = 1$) and the partition function in the general case is given by

$$\Pi_\infty(z) = \left(\frac{\mu - \lambda}{\mu - \lambda z}\right)^{\nu/\lambda} \quad \text{(E.6)}$$

so $\alpha = \nu/\lambda$.

---

[50]If we replace the variable z by $e^{\sqrt{-1}\omega}$ then the characteristic function $\Phi_t(\omega)$ is immediately obtained.

In case $\mu < \lambda$ then $T(t)$ grows progressively larger with time. In all cases, $\Pi_t(1) = 1$ so that probability is conserved. The expressions for the mean and variance of the population remain positive and increase with time, the normalized variance tending to a limiting value, which we read off as

$$\frac{\left(\lambda^2 - \mu^2\right) n + \lambda \nu}{\left((\lambda - \mu) n + \nu\right)^2}. \tag{E.7}$$

Observe that this depends on the initial population n. However, the mean value of the population increases without bound; any value might reasonably be chosen as n at an arbitrarily chosen 'initial' time. The master equation does not establish an equilibrium population, and accordingly, the population is prevented from being asymptotically stationary. Therefore, these values of the underlying population rate parameters can be disregarded as a description of a physical system in equilibrium, such as we have encountered in practice.

Remarks. *On pure birth-death processes.* An intriguing feature emerges in the case of a pure birth-death process when these rates coincide, with respect to the first two moments. From the partition function finite-time exact solution we deduce the following expressions for the mean and variance of the population, at time t,

$$m_t = \exp(\lambda - \mu)t, \tag{E.8}$$

$$\text{var}(t) = \begin{cases} \frac{\lambda+\mu}{\lambda-\mu} e^{(\lambda-\mu)t} \left[e^{(\lambda-\mu)t} - 1\right], & \text{if } \lambda \neq \mu, \\ 2\lambda t & \text{if } \lambda = \mu. \end{cases} \tag{E.9}$$

Observe, therefore, that if $\lambda = \mu$ then as time progresses the mean remains at a constant value of unity and the variance grows without bound. However, it is evident from the exact partition function solution in this case

$$\Pi_t(z) = \frac{1 - (\lambda t - 1)(z - 1)}{1 - \lambda t (z - 1)} \tag{E.10}$$

that the extinction probability, conveniently obtained by setting $z = 0$ in the time-dependent partition function,

$$p_0(t) \equiv \Pi_t(0) \tag{E.11}$$

tends to unity as $t \to \infty$. The population is therefore said to become asymptotically *extinct*.

This fact would seem to be irreconcilable with the properties of the first two moments noted above, namely that the mean population size is a constant unity and that its variance tends to infinity. The resolution of this apparent paradox lies in considering the detailed behaviour of the occupation probabilities $p_N(t)$.

APPENDIX E

Although these probabilities decay to zero with time for all $N \geq 1$, the rate at which they decay *decreases* with time, in such a way that the noted behaviour of the first and second moments is admissible. To see this explicitly, we may extract the $p_N(t)$ from the partition function by writing (E.10) in the form

$$\Pi_t(z) = \left[\frac{\lambda t}{1+\lambda t} + \left(\frac{1-\lambda t}{1+\lambda t}\right)z\right]\left[1 - \left(\frac{\lambda t}{1+\lambda t}\right)z\right]^{-1}. \quad \text{(E.12)}$$

The probabilities $p_N(t)$ can then be obtained as the coefficients of z^N in a binomial series expansion of the above expression for $\Pi_t(z)$, whereby we obtain

$$p_N(t) = \frac{1}{\lambda t(1+\lambda t)}\left(\frac{\lambda t}{1+\lambda t}\right)^N. \quad \text{(E.13)}$$

Thus, with respect to N, the population has a negative exponential decay rate $\log\left(\frac{\lambda t}{1+\lambda t}\right)$ that tends to zero as $t \to \infty$.[51]

The extinction property is also interesting in terms of a sample path of the population process N_t. Given a death rate perfectly counterbalanced by an equal birth rate, one might perhaps intuitively expect the population level to remain stable, in some sense, fluctuating around some positive value. The apparent paradox is resolved by realizing that, at some finite time and with probability 1, the population level will hit $N_t = 0$ and cannot regain any positive value subsequently, because of the absence of immigration; i.e. a situation of population extinction. In other words, although the transition rates are precisely symmetric with respect to increasing and decreasing N_t, the property of extinction is asymmetric in this sense, being represented by the boundary at $N = 0$.

Special cases of the BDI model where a limiting asymptotic distribution exists can also be calculated for (limiting) values of the population parameters. This question is best addressed via analysis of the characteristic or partition function. We explain the behaviour for various values of the parameter $\alpha = \nu/\lambda$.[52]

E.0.1.1 $\alpha = 1$: In this case we obtain the *geometric* distribution, and according to the results of Section 7.3, this yields the exponential distribution in the continuum limit.

E.0.1.2 $\alpha \to 0$: (i) if $\nu \to 0$, with λ, μ fixed, then $\Pi \to 1$ so that the distribution is concentrated at $N = 0$, i.e. extinction . (ii) if $\lambda \to \infty$, with ν fixed (so $\mu \to \infty$ to maintain $\langle N \rangle > 0$); say $\lambda/\mu \to k \leq 1$, then for all values of k the partition function tends to unity, so we have extinction as in (i). A case of interest arises if $\mu = \lambda + 1$ so that $\langle N \rangle = \nu$; nevertheless, $\Pi \to 1$,[53] i.e. asymptotic extinction.

---

[51] The constant mean property $m_t = 1$ can be verified from (E.13) using the series identity $\sum_1^\infty N\zeta^N \equiv \zeta/(1-\zeta)^2$.

[52] Recall that, for a limiting population to exist, $\bar{N} = \nu/(\mu - \lambda)$ for values $\mu > \lambda$.

[53] This can be verified using the fact $\lim_{u \to 0} u^u = 1$.

E.0.1.3 $\alpha \to \infty$: (i) if $\nu \to \infty$ with λ fixed, the Fano factor $\mathcal{F} > 1$ and the average population size $\langle N \rangle \to \infty$; then the partition function $\Pi \to 0$ (for $z \neq 1$) so that the distribution is negative binomial for all values of ν with $p_N \to 0$, $\forall N$. (ii) if $\lambda \to 0$ with ν fixed, then $\mathcal{F} \to 1$ and the limiting distribution is *Poisson* with mean ν/μ.[54]

---

[54] By virtue of the identity $\lim_{s \to 0}(1 + \mu s)^{1/s} = \exp \mu$.

APPENDIX F

SUMMARY OF K-SCATTERING

It is shown in the Rayleigh case of a fixed step number that the amplitude obeys a complex Ornstein–Uhlenbeck equation, and a corresponding stochastic differential equation (SDE) in the K-distributed case is derived.

F.1   Rayleigh scattering

In the Rayleigh case, consider the random walk model for the scattered electric field (cf. Jakeman 1980; Tough 1987; Jakeman and Tough 1988) with step $s^{(j)}$

$$\mathcal{E}_t^{(N)} = \sum_{j=1}^N \overbrace{\exp\left[i\varphi_t^{(j)}\right]}^{s^{(j)}} \tag{F.1}$$

for constant population size N. Since Maxwell's equations for the electromagnetic field possess $U(1)$ gauge invariance with respect to duality rotations, i.e. multiplication by $\exp(i\Lambda)$ for constant Λ (cf. Penrose and Rindler 1984), the assumption of independence of $\{\varphi^{(j)}\}$ implies that these phases are uniformly distributed. Accordingly, in (F.1) the phase factors $\{\exp\left[i\varphi_t^{(j)}\right]\}$ are independent and uniformly distributed on the unit circle in $\mathbf{C}$. Our (phase) diffusion model therefore takes $\{\varphi_t^{(j)}\}$ as a collection of (displaced) Wiener processes on a suitable timescale, $\varphi_t^{(j)} = \Delta^{(j)} + \mathcal{B}^{\frac{1}{2}} W_t^{(j)}$, with the random initializations $\{\Delta^{(j)}\}$ a set of independent random variables uniformly distributed on the interval $[0, 2\pi)$, and thus $d\varphi_t^{(j)} = \mathcal{B}^{\frac{1}{2}} dW_t^{(j)}$, $d\varphi_t^{(j)2} = \mathcal{B} dt$. From Ito's formula (e.g. Oksendal 1998; Karatzas and Shreve 1988) the Ito differential of (F.1) is

$$d\mathcal{E}_t^{(N)} = \sum_{j=1}^N \left(id\varphi_t^{(j)} - \frac{1}{2}d\varphi_t^{(j)2}\right) \exp\left[i\varphi_t^{(j)}\right]. \tag{F.2}$$

The first term $\sum_{j=1}^N id\varphi_t^{(j)} \exp\left[i\varphi_t^{(j)}\right]$ on the right-hand side of (F.2) consists of a sum of independent randomly phased Wiener processes, with variance equal to $\mathcal{B}N dt$, while the second term is independent of the scatterer label j. Thus from (F.2) we can write

$$d\mathcal{E}_t^{(N)} = -\frac{1}{2}\mathcal{B}\mathcal{E}_t^{(N)} dt + (\mathcal{B}N)^{\frac{1}{2}} d\xi_t \tag{F.3}$$

where ξ_t is a complex Wiener process satisfying $|d\xi_t|^2 = dt$, $d\xi_t^2 = 0$. The process ξ_t is adapted to the filtration $\mathcal{F}^{(\varphi)} = \bigcup_j \mathcal{F}^{(j)}$, where $\mathcal{F}^{(j)}$ is the filtration appropriate to the component scatterer phase $\varphi_t^{(j)}$. The (normalized) amplitude process $\mathcal{E}_t$ is then defined by $\mathcal{E}_t = \lim_{N\to\infty}\left[\mathcal{E}_t^{(N)}/\bar{N}^{\frac{1}{2}}\right]$ and satisfies the SDE

$$d\mathcal{E}_t = -\frac{1}{2}\mathcal{B}\mathcal{E}_t dt + (\mathcal{B}x)^{\frac{1}{2}}d\xi_t \qquad (F.4)$$

where the continuous-valued random variable x, the average scattering power, arises from an asymptotically large population via $x = \lim_{N\to\infty}\left[N/\bar{N}\right]$.

F.2 K-distributed noise

In the case of step number fluctuations in the random walk model (F.1), we define the K-amplitude ψ_t as a modification to the Rayleigh amplitude, such that we employ a time-dependent N_t such that $x_t = \lim_{N_t\to\infty}\left[N_t/\bar{N}\right]$. Thus

$$\psi_t = \lim_{N_t\to\infty}\left\{\frac{1}{\bar{N}^{1/2}}\sum_{j=1}^{N_t}\exp\left[i\varphi_t^{(j)}\right]\right\} \qquad (F.5)$$

$$= \lim_{N_t\to\infty}\left\{\left(\frac{N_t}{\bar{N}}\right)^{1/2}\frac{1}{N_t^{1/2}}\sum_{j=1}^{N_t}\exp\left[i\varphi_t^{(j)}\right]\right\} \qquad (F.6)$$

$$= x_t^{\frac{1}{2}}\gamma_t, \qquad (F.7)$$

where $\gamma_t = \lim_{N\to\infty}\left[\mathcal{E}^{(N_t)}/N_t^{\frac{1}{2}}\right]$. Thus the intensity has the compound representation $z_t = x_t u_t$ where $u_t = |\gamma_t|^2$ is the (instantaneous) intensity of the component Rayleigh process. According to the arguments given in the Rayleigh case above, γ_t is a complex Ornstein–Uhlenbeck process which obeys the SDE

$$d\gamma_t = -\frac{1}{2}\mathcal{B}\gamma_t dt + \mathcal{B}^{\frac{1}{2}}d\xi_t. \qquad (F.8)$$

Observe from (F.4), therefore, that γ_t is a unit power Rayleigh process. The above equation for γ_t can be solved by considering the stochastic differential $d\left[\exp(\frac{1}{2}\mathcal{B}t)\gamma_t\right]$, which leads to the solution

$$\gamma_t = \exp\left(-\frac{1}{2}\mathcal{B}t\right)\left\{\gamma_0 + \mathcal{B}^{\frac{1}{2}}\int_0^t \exp\left(\frac{1}{2}\mathcal{B}s\right)d\xi_s\right\}. \qquad (F.9)$$

We deduce the expectation formulae

$$\mathbf{E}[\gamma_t] = \exp\left(-\frac{1}{2}\mathcal{B}t\right)\gamma_0, \qquad (F.10)$$

$$\mathbf{E}\left[|\gamma_t|^2\right] = 1 + \exp(-\mathcal{B}t)(|\gamma_0|^2 - 1). \qquad (F.11)$$

It follows that $\lim_{t\to\infty}\mathbf{E}\left[|\gamma_t|^2\right] = 1$ and so from (F.7) we find the intensity process, defined by $z_t = |\psi_t|^2$, satisfies the conditional expectation property

$\mathbf{E}_{|x}[z_t] = x_t$. The SDE for ψ_t can then be derived by applying the Ito product formula to (F.7). This requires the SDE for the scattering cross-section to be specified. In accordance with the birth–death–immigration (BDI) model (Bartlett 1966), we shall take the re-scaled population variate $x \mapsto \alpha x$ to satisfy the SDE

$$\mathrm{d}x_t = \mathcal{A}(\alpha - x_t)\mathrm{d}t + (2\mathcal{A}x_t)^{\frac{1}{2}}\mathrm{d}W_t^{(x)} \tag{F.12}$$

for an independent Wiener process $W_t^{(x)}$ (Field and Tough 2003a). (In terms of the underlying population parameters of the BDI model, $\alpha = \nu/\lambda$, the ratio of the immigration to birth rate, the birth and death rates coinciding for an infinite-sized population.) Accordingly, x_t has an asymptotic Γ-distribution

$$\Gamma_\alpha(x) = \frac{x^{\alpha-1}\exp(-x)}{\Gamma(\alpha)} \tag{F.13}$$

with first two moments $\mathrm{Var}[x] = \langle x \rangle = \alpha$. These preliminaries enable us to provide the dynamics of the amplitude, intensity, and phase of the K-scattering process as follows (cf. Field and Tough 2003b for a detailed treatment).

F.2.1 Amplitude

Theorem F.1 *The K-amplitude is governed by the SDE*

$$\frac{\mathrm{d}\psi_t}{\psi_t} = \frac{1}{2}\mathcal{B}\mathrm{d}t + \frac{\mathcal{B}^{\frac{1}{2}}}{\gamma_t}\mathrm{d}\xi_t + \mathcal{A}\left(\frac{2(\alpha - x_t) - 1}{4x_t}\right)\mathrm{d}t + \left(\frac{\mathcal{A}}{2x_t}\right)^{\frac{1}{2}}\mathrm{d}W_t^{(x)}. \tag{F.14}$$

This evolution is invariant under the $U(1)$ gauge transformation $\psi_t \mapsto \exp(i\Lambda)\psi_t$, for constant Λ.

F.2.2 Intensity

Proposition F.2 *The K-intensity SDE is given by*

$$\mathrm{d}z_t = \left[\mathcal{B}(x_t - z_t) + \frac{\mathcal{A}z_t(\alpha - x_t)}{x_t}\right]\mathrm{d}t + \left(2\mathcal{B}x_t z_t + \frac{2\mathcal{A}z_t^2}{x_t}\right)^{\frac{1}{2}}\mathrm{d}W_t^{(z)} \tag{F.15}$$

in which $W_t^{(z)}$ is correlated with $W_t^{(x)}$ of (F.12), and satisfies

$$\left(2\mathcal{B}x_t z_t + \frac{2\mathcal{A}z_t^2}{x_t}\right)^{\frac{1}{2}}\mathrm{d}W_t^{(z)} = (2\mathcal{B}x_t z_t)^{\frac{1}{2}}\mathrm{d}W_t^{(r)} + \left(\frac{2\mathcal{A}}{x_t}\right)^{\frac{1}{2}}z_t\mathrm{d}W_t^{(x)} \tag{F.16}$$

and $W_t^{(r)}$ is a real-valued Wiener process defined by

$$\gamma_t^*\mathrm{d}\xi_t + \gamma_t\mathrm{d}\xi_t^* \equiv \left(\frac{2z_t}{x_t}\right)^{\frac{1}{2}}\mathrm{d}W_t^{(r)}. \tag{F.17}$$

F.2.3 Phase

Proposition F.3 *The resultant phase θ_t of the K-amplitude process satisfies the SDE*

$$d\theta_t = \left(\frac{\mathcal{B}x_t}{2z_t}\right)^{\frac{1}{2}} dW_t^{(\theta)} \qquad (F.18)$$

where the distinct (real-valued) Wiener process $W_t^{(\theta)}$ is defined according to

$$\frac{1}{i}(\gamma_t^* d\xi_t - \gamma_t d\xi_t^*) \equiv \left(\frac{2z_t}{x_t}\right)^{\frac{1}{2}} dW_t^{(\theta)}. \qquad (F.19)$$

These various relations allow the K-amplitude dynamics to be recast in terms of $W^{(x)}$, $W^{(\theta)}$, $W^{(r)}$ as follows.

Corollary F.4 *The K-amplitude satisfies the SDE*

$$\frac{d\psi_t}{\psi_t} = \left[\mathcal{A}\left(\frac{2(\alpha - x_t) - 1}{4x_t}\right) - \frac{1}{2}\mathcal{B}\right] dt + \left(\frac{\mathcal{A}z_t + \mathcal{B}x_t^2}{2x_t z_t}\right)^{\frac{1}{2}} dW_t^{(z)} + i\left(\frac{\mathcal{B}x_t}{2z_t}\right)^{\frac{1}{2}} dW_t^{(\theta)} \qquad (F.20)$$

in which, alternatively, the Wiener terms can be expressed as

$$\left(\frac{\mathcal{A}}{2x_t}\right)^{\frac{1}{2}} dW_t^{(x)} + \left(\frac{\mathcal{B}x_t}{2z_t}\right)^{\frac{1}{2}} \left(dW_t^{(r)} + i dW_t^{(\theta)}\right). \qquad (F.21)$$

The following result, implied by (F.14) and the identities $d\xi_t^2 = d\xi_t dW_t^{(x)} = 0$, is relevant in connection with the geometry of fluctuations for weak scattering processes discussed in Section 9.3.

Corollary F.5 *The product cross-section/K-amplitude stochastic differentials satisfy*

$$dx_t^2 = 2\mathcal{A}x_t dt, \qquad (F.22)$$

$$dx_t d\psi_t = \mathcal{A}\psi_t dt, \qquad (F.23)$$

$$d\psi_t^2 = \left(\frac{\mathcal{A}\psi_t^2}{2x_t}\right) dt, \qquad (F.24)$$

$$|d\psi_t|^2 = \left(\frac{\mathcal{A}z_t}{2x_t} + \mathcal{B}x_t\right) dt. \qquad (F.25)$$

F.2.4 Geometry of fluctuations

We observe from Propositions F.2 and F.3 that $dW_t^{(z)} dW_t^{(\theta)} = 0$, so the fluctuations in R_t, θ_t are statistically independent. The relative magnitude of the radial and orthogonal (phase) fluctuations is determined by

$$\frac{\Sigma_t^{(z)}}{\Sigma_t^{(\theta)}} = 4z_t^2 + \frac{4\mathcal{A}z_t^3}{\mathcal{B}x_t^2} \tag{F.26}$$

which exceeds the quotient obtained in the Rayleigh case, $\mathcal{A} = 0$. These relations can be used to characterize the geometry of the K-scattering amplitude fluctuations as follows. The real and imaginary parts of the resultant amplitude I, Q are the usual and quadrature-phase components, respectively.

Proposition F.6 *In the K-distributed case, $A \neq 0$, the amplitude diffusion tensor is non-degenerate, and the fluctuations in the and quadrature phase components δI_t, δQ_t are correlated. The (comoving) error surface $\mathcal{S}$ of $\delta\psi_t$ is an ellipse whose major axis lies in the instantaneous radial direction defined by ψ_t. Degeneracy occurs in the Rayleigh case $\mathcal{A} = 0$, for which $\mathcal{S}$ is a circle, i.e. the fluctuations in ψ_t are isotropic.*

A complete account of the dynamical properties of K-scattering is given in Chapter 8 (*orig.* Field and Tough 2003*b*).

APPENDIX G

ITERATIVE SOLUTION FOR VECTOR PROCESSES

The generalization to a vector process is straightforward. We consider the vector **x** with components labeled with Greek superscripts, i.e. x^α; the drift **F** is also a vector. The Brownian process is a vector also and we label its components with Greek superscripts, while the time variable appears as a Latin subscript, thus W_t^α. Finally, the volatility is now a matrix represented by σ_β^α. Derivatives with respect to the components of **x** are denoted in standard tensor notation as follows

$$F^\alpha_{,\beta} = \frac{\partial F^\alpha}{\partial x^\beta}; \quad \sigma^\alpha_{\beta,\gamma} = \frac{\partial \sigma^\alpha_\beta}{\partial x^\gamma}. \tag{G.1}$$

We then have

$$dx^\alpha = F^\alpha(\mathbf{x}(t))\,dt + \sigma^\alpha_\beta(\mathbf{x}(t))\,dW_t^\beta \tag{G.2}$$

with summation on repeated Greek indices, according to the Einstein summation convention introduced in Chapter 3. The corresponding integral equation is

$$\Delta x^\alpha = \int_0^t F^\alpha(\mathbf{x}(u))\,du + \int_0^t \sigma^\alpha_\beta(\mathbf{x}(u))\,dW_u^\beta. \tag{G.3}$$

The iterative solution to this integral equation can be developed exactly as it was in the scalar case. To second order in time we find this to be, having taken care to place terms in the order corresponding to the scalar solution (11.16),

$$\Delta x^\alpha = tF^\alpha + \sigma^\alpha_\beta W_t^\beta + \frac{1}{2}t^2 F^\gamma F^\alpha_{,\gamma} + \sigma^\mu_\tau F^\alpha_{,\mu}\int_0^t W_s^\tau\,ds$$

$$+ F^\mu \sigma^\alpha_{\beta,\mu}\int_0^t s\,dW_s^\beta + \sigma^\mu_\tau \sigma^\alpha_{\beta,\mu}\int_0^t dW_s^\beta W_s^\tau + \frac{1}{2}\sigma^\mu_\tau \sigma^\nu_\sigma F^\alpha_{,\mu\nu}\int_0^t W_s^\tau W_s^\sigma\,ds$$

$$+ \frac{1}{2}\sigma^\mu_\tau \sigma^\nu_\sigma \sigma^\lambda_{\gamma,\mu\nu}\sigma^\alpha_{\beta,\lambda}\int_0^t dW_s^\beta \int_0^s dW_u^\nu W_u^\tau W_u^\sigma + \sigma^\mu_\tau F^\nu \sigma^\alpha_{\beta,\mu\nu}\int_0^t s W_s^\tau\,dW_s^\beta$$

$$+ \sigma_\tau^\mu \sigma_{\sigma,\mu}^\nu F_{,\nu}^\alpha \int_0^t ds \int_0^s dW_u^\sigma W_u^\tau + \sigma_\tau^\mu \sigma_\sigma^\nu \sigma_{\gamma,\nu}^\lambda \sigma_{\beta,\mu\lambda}^\alpha \int_0^t dW_s^\beta W_s^\tau \int_0^s dW_u^\gamma W_u^\sigma$$

$$+ \frac{1}{6} \sigma_\tau^\mu \sigma_\sigma^\nu \sigma_\gamma^\lambda \sigma_{\beta,\mu\nu\lambda}^\alpha \int_0^t dW_s^\beta W_s^\sigma W_s^\tau W_s^\gamma + \sigma_\tau^\mu \sigma_{\sigma,\mu}^\nu \sigma_{\beta,\nu}^\alpha \int_0^t dW_s^\beta \int_0^s dW_u^\sigma W_u^\tau$$

$$+ \sigma_\tau^\mu F_{,\mu}^\nu \sigma_{\beta,\nu}^\alpha \int_0^t dW_s^\beta \int_0^s W_u^\tau du + F^\mu \sigma_{\tau,\mu}^\nu \sigma_{\beta,\lambda}^\alpha \int_0^t dW_s^\beta \int_0^s u dW_u^\tau$$

$$+ \frac{1}{2} \sigma_\tau^\mu \sigma_\gamma^\nu \sigma_{\beta,\mu\nu}^\alpha \int_0^t dW_s^\beta W_s^\tau W_s^\gamma$$

$$+ \sigma_\tau^\mu \sigma_{\sigma,\mu}^\nu \sigma_{\gamma,\nu}^\lambda \sigma_{\beta,\lambda}^\alpha \int_0^t dW_s^\beta \int_0^s dW_u^\gamma \int_0^u dW_v^\sigma W_v^\tau. \qquad (G.4)$$

The calculation of the expectation values for the increments and their products proceeds in much the same way as the scalar case; some care is needed in the ordering of the various indices. Thus we write the stochastic part of the vector increment as

$$\Delta \hat{x}^\alpha = \Delta x^\alpha - t F^\alpha - \frac{1}{2} t^2 F^\gamma F_{,\gamma}^\alpha. \qquad (G.5)$$

In a similar way to the scalar case, we have

$$\mathbf{E}\left(\Delta \hat{x}^\alpha\right) = \frac{t^2}{4} \sigma_\tau^\mu \sigma_\tau^\nu F_{,\mu\nu}^\alpha \qquad (G.6)$$

while the covariance matrix is given by

$$\mathbf{E}\left(\Delta \hat{x}^\alpha \Delta \hat{x}^\lambda\right) = t \sigma_\beta^\alpha \sigma_\beta^\lambda + \frac{t^2}{2} \left(\sigma_\gamma^\lambda \sigma_\gamma^\beta F_{,\beta}^\alpha + \sigma_\gamma^\alpha \sigma_\gamma^\beta F_{,\beta}^\lambda\right) + \frac{t^2}{2} \left(\sigma_\gamma^\lambda F^\beta \sigma_{\gamma,\beta}^\alpha + \sigma_\gamma^\alpha F^\beta \sigma_{\gamma,\beta}^\lambda\right)$$

$$+ \frac{t^2}{4} \left(\sigma_\gamma^\lambda \sigma_\beta^\sigma \sigma_\beta^\mu \sigma_{\gamma,\sigma\mu}^\alpha + \sigma_\gamma^\alpha \sigma_\beta^\sigma \sigma_\beta^\mu \sigma_{\gamma,\sigma\mu}^\lambda\right) + \frac{t^2}{2} \sigma_\gamma^\beta \sigma_\gamma^\sigma \sigma_{\nu,\beta}^\alpha \sigma_{\nu,\sigma}^\lambda. \qquad (G.7)$$

Extension of these results beyond second order in time in the general multiplicative noise case, although proceeding along similar lines, is rather computationally heavy and is not considered here.

APPENDIX H

OPEN PROBLEMS

1. Construction of scattering dynamics for populations arising from higher order transitions and/or non-linearity in the generation and recombination coefficients. Investigation of the continuum limit therein (in relation to the Pawula theorem).

2. Extension of the local population model to a spatially correlated one, induced by the effect of inter-site migration, on a discrete lattice of sites and the continuum limit thereof, in relation to path integrals. The projection of such spatial model to single site and the relationship with the BDI model.

3. Optimal inference of the scattering cross-section for given sample rate through adjustment of smoothing parameters, for K-scattering processes, and the extension to more general population dynamics.

4. Filtering of additive noise in superposition with the scattered amplitude process for K-scattering and weak scattering processes, and the extension to more general population models.

5. Coherent signal extraction for weak scattering processes, de-noising, and the systematic analysis of performance and error bounds for such procedures.

6. Estimation of the scattering cross-section for weak scattering models through analysis of fluctuations, as already achieved for the K-scattering model, to include general populations.

7. Formulation of a stochastic volatility based anomaly detection scheme for weak scattering processes, as an extension of the corresponding results achieved for K-scattering.

8. Incorporation of the stochastic dynamical model for radar scattering into a Bayesian detection/tracking algorithm for sea clutter environments and performance comparison with the stochastic volatility based anomaly detection scheme (e.g. via computation of the associated received operator characteristic curves).

9. To investigate the possibility of chaotic behaviour, in the deterministic non-linear evolution of the 'noise free skeleton' of the coupled amplitude/cross-section dynamics.

10. To explore the fundamental physical limits that exist with respect to the observability of the scattering cross-section from the intensity and phase fluctuations, in a variety of contexts.

11. To establish the relationship between the z-parameter quantifier of nonlinearity and the shape parameter of the SDE radar scattering model, from a theoretical point of view; to exhibit such relationship by evaluating the z-parameter on a data set synthetically generated from the SDE model.

APPENDIX I

SUGGESTED FURTHER READING

[A] A. Ishimaru, *Wave Propagation and Scattering in Random Media*, Volume I: *Single scattering and transport theory*; Volume II: *Multiple scattering, turbulence, rough surfaces and remote sensing*, Academic Press, New York, 1978. (Translated into Russian, 1981 and Chinese, 1986.) IEEE Press-Oxford University Press Classic Re-issue, 1997.

[B] A. Ishimaru, *Electromagnetic Wave Propagation, Radiation, and Scattering*, Prentice Hall, New Jersey, 1991.

[C] Mauro Fabrizio, Angelo Morra, and Angelo Morro, *Electromagnetism of Continuous Media: Mathematical Modelling and Applications*, Oxford University Press, July 2003, ISBN: 0198527004.

[D] Eric Jakeman and Kevin D. Ridley *Modeling Fluctuations in Scattered Waves*, Taylor and Francis, June 19, 2006. ISBN: 0750310057.

[E] Keith D. Ward, Robert J.A. Tough, and Simon Watts *Sea Clutter: Scattering, the K-distribution and Radar Performance*, IET Press, June 16, 2006, ISBN: 0-86341-503-2, 978-086341-503-6.

Items [A, B] are important texts by a leading international expert in the field, on the subject of electromagnetic scattering from random media. However, there is no coverage of the approach using stochastic calculus, or of the K-scattering process and its extensions to weak scattering models.

Item [C] deals with wave propagation in continuous electromagnetic media including effects of non-locality and 'memory' – no coverage of scattering from random media. Complements the proposed monograph well. An alternative recent addition to the literature on the subject, item [D], should be an important text by a pioneer in the field of non-Gaussian models of scattered radiation. Mathematically, it is less sophisticated than the present monograph, covering basic statistical models, Gaussian, Gamma, K-lognormal etc., random walks, phase screen scattering, propagation through extended media, multiple scattering, vector wave scattering, K-distributed noise, experimental artifacts, and numerical simulation. The book is aimed at the applied physics/systems engineering user and thus complements the present more theoretical treatment well. Indeed one chapter describes K-distributed noise and includes mention of some of the recent work of the current author in the journal literature, in abbreviated and simplified form.

The most recent addition to the literature in book form, item [E], is devoted to the study of sea clutter: scattering, the K-distribution and radar performance.

APPENDIX I

The text gives an authoritative account of our current understanding of radar sea clutter from an engineering perspective. The authors pay particular attention to the compound K-distribution model, which they have helped develop over the past two decades. Evidence supporting this model, including a detailed review of the calculation of EM scattering by the sea surface, its statistical formulation, and practical application to the specification, design and evaluation of radar systems are included. The calculation of the performance of practical radar systems is emphasized. This book provides a less mathematically sophisticated treatment in the radar context which will serve as an excellent complement to the current monograph for specialist radar engineers, and also be of interest to the wider applied physics and mathematics academic communities.

REFERENCES

Abramowitz, M. and Stegun, I. A. (Eds.) (1970). *Handbook of Mathematical Functions*. Dover, New York.

Bachelier, L. (1900). Théorie de la spéculation, *Annales Scientifiques de l'École Normale Supérieure* **3** (17), 21–86.

Bartlett, M. S. (1966). *An Introduction to Stochastic Processes*, ch. 3, Cambridge University Press.

Batchelor, G. K. (1967). *An Introduction to Fluid Dynamics*, Cambridge University Press.

Copson, E. T. (1970). *An Introduction to the Theory of Functions of a Complex Variable*, Section 10.63. Clarendon Press, Oxford.

Einstein, A. (1905). On the movement of small particles suspended in a stationary liquid demanded by the molecular-kinetic theory of heat. Re-printed in *Investigations on the Theory of the Brownian Movement*, A. Einstein (Author), R. Fürth (Editor), A. D. Cowper (Translator) Publisher: Dover (1956). ASIN: B000F62YS0.

Ermak, D. L. and McCammon, J. A. (1978). *J. Chem. Phys.* **59**, 1352.

Ernst, R. R., Bodenhausen, G. and Wokaun, A. (1987). *Principles of Nuclear Magnetic Resonance in One and Two Dimensions*, Oxford: Clarendon Press.

Fano, U. (1947). *Phys. Rev.* **72**, 26–29.

Fayard, P. (2008). *Private communication*.

Fayard, P. and Field, T. R. (2008). Optimal inference of the scattering cross-section through the phase decoherence, *Waves in Random and Complex Media*, in press.

Feng T., Field, T. R. and Haykin, S. (2007). Stochastic differential equation theory applied to wireless channels, *IEEE Transactions on Communications*, **55** (8), 1478–1483.

Feng, T. and Field, T. R. (2008). Statistical analysis of mobile radio reception – an extension of Clarke's model, *IEEE Transactions on Communications*, in press.

Field, T. R. (2002). *Stochastic differential equations and their application to the characterization of sea clutter*, Invited Lectures, October 7–8, Adaptive Systems Laboratory, McMaster University.

Field, T. R. (2003). Quantum diffusion on manifolds. *J. Geom. Phys.* **47**, 484–496.

Field, T. R. (2005). Observability of the scattering cross-section through phase decoherence. *J. Math. Phys.* **46**, 063305.

Field, T. R. (2006). Spin diffusion – a new perspective in magnetic resonance imaging. Ch. 5 in *New Directions in Statistical Signal Processing: From Systems to Brain*, MIT Press.

Field, T. R. and Bain, A. D. (2008). Origins of stochastic spin noise dynamics and implications for experimental NMR, *Phys. Rev. E*, submitted; 49th ENC, March 9–14, 2008, Asilomar Conference Grounds, Pacific Grove, California.

Field, T. R. and Haykin, S. (2008). Non-linear dynamics of sea clutter. *International Journal on Navigation and Observation*, Special Issue on Modelling and Processing of Radar Signals for Earth Observation; M. Greco, S. Watts (eds.), forthcoming.

Field, T. R. and Tough, R. J. A. (2003a). Diffusion processes in electromagnetic scattering generating K-distributed noise, *Proc. R. Soc. Lond. A* **459**, 2169–2193.

Field, T. R. and Tough, R. J. A. (2003b). Stochastic dynamics of the scattering amplitude generating K-distributed noise, *J. Math. Phys.* **44** (11), 5212–5223.

Field, T. R. and Tough, R. J. A. (2005). Dynamical models of weak scattering. *J. Math. Phys.* **46** (1), 13302–13320.

Fox, L. and Mayers, D. F. (1968). *Computing methods for scientists and engineers*. Chapter 10, Clarendon Press, Oxford.

Gini, F. and Greco, M. (2001). Texture modeling and validation using recorded high resolution sea clutter data. *Proc. 2001 IEEE Radar Conference*, Atlanta, GA, May 1–3 2001, pp. 387–392.

Greco, M. and Gini, F. (2007). Sea clutter nonstationarity: the influence of long waves. In S. Haykin (ed.), *Adaptive Radar Signal Processing*, Wiley; Chapter 5, pp. 159–191.

Greenside, H. S. and Helfand, E. (1981). Numerical integration of stochastic differential equations – II, *Bell Syst. Tech. J.*, **60**, 1927–1940.

Haykin, S. (1999). Radar clutter attractor: implications for physics, signal processing, and control. *IEE Proceedings on Radar, Sonar and Navigation, Vol. 146, No. 4, pp. 177*, August, Invited.

Haykin, S. (2001). *Communication Systems*, 4th edn, Wiley (New York).

Haykin, S. (2006) (ed.) *Adaptive Radar Signal Processing*, Wiley-Interscience ISBN-10: 0471735825; Chapter 4.

Haykin, S., Bakker, R. and Currie B. W. (2002). Uncovering nonlinear dynamics – The case study of sea clutter. *Proc IEEE* **90** (5), 860–881.

Haykin, S. and Thomson, D. J. (1998). Signal detection in a nonstationary environment reformulated as an adaptive pattern classification problem. *Proc. IEEE*, Vol. 86, pp. 2325–2344.

Helstrom, C. W. (1960). *Statistical Theory of Signal Detection*, Pergamon, Oxford.

Higham, D. J. (2004). *An Introduction to Financial Option Valuation: Mathematics, Stochastics and Computation*, Cambridge University Press.

Howison, S. (2004). *Private communication*.

Hughes, A. J., Jakeman, E., Oliver C. J. and Pike, E. R. (1973). Photon correlation spectroscopy; dependence of linewidth error on normalisation, clip level detector area, sample time and count rate. *J. Phys. A, Math. and Gen.* **6**, 1327.

Hughston, L. P. (1996). Geometry of stochastic state vector reduction *Proc. R. Soc. London*, Ser. A **452**, 953.

Jakeman, E. (1980). On the statistics of K-distributed noise. *J. Phys. A* **13**, 31-48.

Jakeman, E., Hopcraft, K. I. and Matthews, J. O. (2003). Distinguishing population processes by external monitoring. *Proc. R. Soc. Lond.* A, **459**, 623–639.

Jakeman, E. and Tough, R. J. A. (1987). The generalised K-distribution: a statistical model for weak scattering. *J. Opt. Soc. Am.*, **A4**, 1764.

Jakeman, E. and Tough, R. J. A. (1988). Non-Gaussian models for the statistics of scattered waves. *Adv. Phys.* **37**, 471.

Jakeman, E., Watson, S. M. and Ridley, K. D. (2001). Intensity-weighted phase-derivative statistics. *J. Opt. Soc. Am.* A **18** (9), 2121–2131.

Jeffreys, H. and Jeffreys, B. S. (1966). *Methods of Mathematical Physics*, 3rd Edn., Cambridge University Press.

Karatzas, I. and Shreve, S. E. (1988). *Brownian Motion and Stochastic Calculus*. Berlin: Springer.

Klauder, J. R. and Petersen, W. P. (1985). Numerical integration of multiplicative-noise stochastic differential equations. *SIAM J. Numerical Analysis* **22**, 1153–1166.

Kramers, H. A. (1940). *Physica* **7**, 284.

Luttrell, S. P. (2001). *Private communication.*

Moyal, J. E. (1949). *J. Roy. Statist. Soc.* **11**, 1358.

Nelson, E. (1967a). *Dynamical Theories of Brownian Motion*, Princeton University Press.

Nelson, E. (1967b). *Tensor Analysis*, Princeton University Press.

Nelson, E. (1985). *Quantum Fluctuations*, Princeton University Press.

Oksendal, B. (1998). *Stochastic Differential Equations – An Introduction with Applications*, 5th Edition, Springer.

O'Loghlen, J. W. (2001). *Private communication.*

Penrose, R. and Rindler, W. (1984). *Spinors and Space-time*, Vol. 1, Cambridge University Press.

Pusey, P. N. and Tough, R. J. A. (1982). Langevin approach to the dynamics of interacting Brownian particles. *J. Phys* A, **15**, 1291–1308.

Quiang, J. and Habib, S. (2000). A second order stochastic leapfrog algorithm for multiplicative noise Brownian motion. *arXiv:physics/9912055v2*, 4th Jan. 2000.

Rice, S. O. (1954). Mathematical analysis of random noise. *Bell. Syst. Tech. J.*, **23**, **24**, reprinted in *Selected Papers on Noise and Stochastic Processes*, N. Wax, Ed., Dover, New York.

Ridley, K. D., Watson, S. M., Jakeman, E. and Harris, M. (2002). Heterodyne measurements of laser light scattering by a turbulent phase screen. *Applied Optics* **41**, No. 3, 532–542.

Risken, H. (1989). *The Fokker–Planck Equation*, 2nd edn., Springer.

Siegel, A. (1956). *Non-parametric Statistics for the Behavioral Sciences*, McGraw–Hill (New York).

Stone, L. (1992). Coloured noise or low-dimensional chaos? *Proc. R. Soc. Lond.* B **250**, 77–81.

Sugihara, G. (1994). Nonlinear forecasting for the classification of natural time series. *Phil. Trans. R. Soc. Lond.* A **348**, 477–495.

Sugihara, G. and May, R. (1990). Nonlinear forecasting as a way of distinguishing chaos from measurement error in a data series. *Nature, Lond.* **344**, 734–741.

Tough, R. J. A. (1987). A Fokker–Planck description of K-distributed noise. *J. Phys. A* **20**, 551–567.

Tough, R. J. A. (1991). Interferometric detection of sea surface features. *R.S.R.E. Memorandum 4446*.

Van Kampen, N. G. (1961). Power series expansion of the master equation, *Can. J. Phys.* **39**, 551–567.

Van Kampen, N. G. (1981). *Stochastic Processes in Physics and Chemistry*. North Holland, Amsterdam.

Ward, K. D. (1981). Compound representation of high resolution sea clutter. *Electronics Letters* **17**, 561.

Ward, K. D. (2002). *Private communication*.

Whittaker, E. T. and Watson, G. N. (1969). *A Course of Modern Analysis*, Chapter XI, Cambridge University Press.

Wolfram, S. (1999). *The Mathematica Book*, Cambridge University Press, 4th edition.

Wong, E. (1963). *Proc. Am. Math. Soc. Symp. Appl. Math.* **16**, 264.

Wong, E. and Zakai, M. (1965). On the relation between ordinary and stochastic differential equations. *Int. J. Engng Sci.* Vol. 3, pp. 213–229, Pergamon Press.

Wootters, W. K. (1981). Statistical distance and Hilbert space, *Phys. Rev. D* **23**, pp. 357–362.

INDEX

amplitude, 16
  K-, 53
  weak scattering, 70
amplitude fluctuations, 77, 126, 169
  K-scattering, 59, 74
  weak scattering, 74
  geometry of, 26, 59, 74
anomaly detection, 25, 27, 34, 52, 55, 67, 69, 79, 86, 87, 97, 111, 112, 127, 132–134, 148
asymptotic, 25
  distribution, 25, 27, 29, 30, 35, 37, 39, 44, 47, 49, 54, 59–61, 64, 69, 70, 79, 81, 83, 84, 86, 90, 161, 163, 167
  stationarity, 162
asymptotic
  distribution, 41
asymptotic behaviour
  weak scattering, 79
autocorrelation, 5, 52, 53, 63–67, 142
  of Wiener process, 5

Bachelier
  Louis, 3
Bessel
  function, 28, 63, 79, 107, 119, 122
  process, 21, 22, 27
birth-death-immigration (BDI), 46, 47
  continuum diffusion limit of, 48
Black–Scholes
  theory of option pricing, 21, 28, 34, 160
Brownian motion, 3, 5
  and Wiener process, 5, 7, 9
  geometric, 21

Cauchy distribution, 67, 154
  stability of, 66
central limit theorem, 6, 28, 53, 90, 91, 154
chain rule
  for Stratonovich calculus, 156
Christoffel symbols, 14, 15, 156
clutter, 29, 113, 114, 127
  target returns in, 114
coherent, 25, 43, 69, 70, 78, 86, 87, 114, 115, 119
  amplitude, 40

clutter, 112, 113, 127
  radar, 127, 138, 140
  scattering, 52
  superposition, 25, 40, 68, 70
compound model, 90, 91
  for K-scattering, 66, 67, 109
  for weak scattering, 80–83
contravariant
  diffusion tensor, 12, 13, 60
correlation, 20
  function, 35, 38, 40, 42, 64, 67, 111, 112, 122, 127
  standard statistical measure of, 124
  timescales, 31, 35, 55, 58, 108, 143, 146
covariant, 14
  derivative, 14, 16
  Fokker–Planck equation, 14, 18, 60, 62
cross-section
  dynamics, 30
    as limit of BDI process, 48, 49
  generalized, 91
  estimation of, 150, 151
  instantaneous observability of, 150
  radar, 66, 138, 139, 141
  scattering, 20, 25, 26, 28–30, 35, 38, 46, 52, 53, 55, 61, 67, 88, 93, 130, 131, 133, 134, 137, 144, 150–152, 167
current
  stochastic, 5, 12, 17, 18, 31, 35, 36, 38, 39, 60, 62, 72, 85, 86

detailed balance, 12, 18
  in K-scattering models, 62
  in weak scattering models, 85
deterministic dynamics, 42, 108, 120, 126, 143, 148
diffusion
  coefficient, 4, 17, 26, 39, 43, 72, 77
  kinematics of, 16
  on manifold, 12
  process, 12, 13, 17, 18, 25, 27, 29, 31, 33, 43, 68, 69, 85, 152
  quantum mechanical, 18
  requirements of, 48–51
  tensor, 11–16, 19, 35, 38, 59–61, 73, 74, 76, 169

INDEX

Dirac
  delta function, 5, 9
distribution
  K-, 25, 27–30, 35, 52–54, 60, 62, 64, 66, 67, 69, 70, 86, 88, 91, 97, 124–127, 130–132, 134, 137, 145, 154, 165, 166, 169
  asymptotic, 25–30, 35, 37–39, 41, 44, 47, 49, 54, 59–61, 64, 69, 70, 79, 81, 83, 84, 86, 90, 161, 163, 167
  Cauchy, 66, 67, 154
  gamma, 29, 49, 154
  infinitely divisible, 153
  lognormal, 21
  Rayleigh, 27–30, 91
  Rice, 70, 72, 77, 79
  stable, 44, 66, 153, 154
Doppler, 25, 64, 86, 95, 97, 99, 110–115, 118, 144, 152
  effect on volatility analysis, 111
drift
  backward, 12–14, 18
  forward, 12–14, 17, 18, 36, 39

Einstein
  Albert, 3
  Annus Mirabilis, 5
  derivation of heat equation, 3, 4, 6, 17
  summation convention, 12, 16, 26, 170
endogenous model, 67, 112, 115
  defined, 56
ensemble
  average, 3, 20, 27, 33, 34, 36, 125, 134, 143, 147, 159, 160
equation of continuity, 14, 35, 62
expectation
  conditional, 158, 166
experimental tests, 124
extinction, 162, 163

Fano factor, 47, 164
Faraday's law, 152
Feng
  Tao (Stephen), 133
Field
  equations, 54
filtration, 44, 56, 91, 158, 165
Fisher information
  discrimination, 132
  geodesic distance, 132
Fokker–Planck
  equation (FPE), 12–14, 18, 30–32, 35, 36, 41, 42, 47, 49, 60, 61, 63, 66, 72, 86, 99, 106, 107, 111, 119, 148
  covariant, 14, 18, 60, 62
  relation to Kramers–Moyal expansion, 47, 50

gamma
  distribution, 29, 49, 154
  function, 49
  variate, 121, 123, 150
generalized K-scattering, 26, 69–72, 74, 76, 78, 79, 81, 82, 84–86
Girsanov theorem, 28, 33, 34, 159, 160
  relevance to intensity scattering, 28, 33
  role in mathematical finance, 21, 28, 34, 160
Green's function, 42, 45, 53, 63
  solution to heat equation, 5, 17

Haykin
  Simon, 138
heat equation, 1, 3–5, 17, 23
homodyned
  K-scattering, 26, 69, 70, 73, 77, 78, 80–83, 86
hypergeometric function, 65, 66, 83, 122

in-phase component, 40, 143, 152, 169
increments, 59, 60, 76, 108, 109, 119, 121, 129, 131, 132, 171
  independence of, 5–7, 9, 124
infinite divisibility, 153, 154
  relation to stability, 153, 154
intensity, 41, 55
  analysis of fluctuations in, 125, 126, 128
  autocorrelation, 63
iterative solution, 99, 103, 108, 170
  for vector processes, 170
  of SDEs, 105, 108, 109, 170
Ito
  differential, 9
  formula, 10
  isometry, 9
  product rule, 10
  stochastic integral, 7
Ito calculus, 7

K-distribution, 25, 27–30, 35, 52–54, 60, 62, 64, 66, 67, 69, 70, 86, 88, 91, 97, 124–127, 130–132, 134, 137, 145, 154, 165, 166
  infinite divisibility of, 154
K-scattering, 25, 26, 30, 31, 37–40, 42, 52, 59, 60, 68–70, 76, 79, 85–88, 91, 93, 97, 99, 107–111, 145, 154, 165, 167, 169
  compound model, 66, 67, 109
  iterative schemes for, 99

INDEX

Kolmogoroff
  forward drift, 13, 15, 156
Kramers–Moyal
  coefficients, 47, 50
  expansion, 46, 48, 50

Lagrangian, 18
Langevin
  equation, 99, 108, 142
Laplace
  transform variable, 29
Laplace's equation, 23
  stochastic solution to, 23
Levi-Civita
  connection, 12–14, 19
  connection, 16
Lorentzian spectrum, 66, 67, 154

manifold
  diffusion on, 12
Mann–Whitney
  rank-sum statistic, 138, 140, 147, 148
Markov, 42, 52, 59, 121, 158
  non-, 125, 137
martingale, 34, 156, 159
  property of Ito stochastic integrals, 156
master equation, 31, 46, 49, 50, 54, 162
master equation (transitions)
  higher order, 50
mathematical finance, 100, 117, 160
Maxwell's equations, 43, 144, 145, 165
  for time dependent magnetic field, 152
metric
  Euclidean, 13, 14, 16, 22
  role in construction of diffusion, 12, 14
molecules
  random collisions of, 3

Nelson
  Edward, 19
non-linearity, 138–140, 147, 149
  of sea clutter, 138, 147, 148
  shape parameter and, 138, 139, 143, 145–149
numerical solution, 108, 157

optical scattering, 125
  volatility analysis, 125
Ornstein–Uhlenbeck process, 23, 40–42, 44, 45, 53, 60, 82, 89, 109–111, 130, 165, 166
osmotic current / velocity, 5, 17
osmotic equation, 17, 18

partial differential equations, 23
  solution via diffusion process, 23

partition function, 51, 161
Pawula theorem, 47, 51, 172
phase
  decoherence, 89, 151, 152
  dynamics, 43, 93
  wrapping, 44, 45, 94
phasor, 43–45, 89, 94
  random, 10, 39, 45, 89, 90, 94, 143
Poisson
  distribution, 49
  equation, 23
  variate, 161
Poisson process, 49
population
  birth-death-immigration (BDI) process, 47, 161
  dynamics, 46, 48, 88, 91, 92
power, 118
  spectral density, viii, 52, 53, 64–67
    for non-stationary process, 65
power (scattering), 20, 44, 80, 82, 84, 99, 112, 140, 146, 166
propagator, viii, 45, 63–65, 106, 107, 119, 121, 122, 148

quadratic variation, 4, 9, 10
quadrature components, 40, 132
quadrature-phase component, 40, 143, 152, 169

radar
  cross-section (RCS), 66, 138, 139, 141–148
  parameters, 144
  sea clutter, 126, 127
Radon–Nikodym derivative, 34, 159
random phase screen, 125
random walk
  discrete, 6
  extended model, 88, 89, 139
  model of scattering, 26, 35, 37, 38, 40, 42–44, 52, 53, 67–70, 86, 88–91, 94, 129–131, 139, 142, 143, 165, 166
  step number fluctuations, 40, 43, 52, 53, 68, 88, 166
Rice
  distribution, 79
  scattering, 71, 79, 80, 83, 86
Runge–Kutta
  algorithm, 108, 109, 119

scattering
  K-, 52
  optical wavelength, 125
  radar wavelength, 126

184 INDEX

scattering (*Cont.*)
  Rayleigh, 40
  weak, 69
sea clutter, 127
  z-parameter and, 140
sea state, 138–140, 145–148
  z-parameter and, 140, 147, 149
second order algorithms, 97, 99, 115
shape parameter, 29, 80, 84
simulation, 99, 150
spectral density, viii, 52, 53, 64–67
stability, 22, 44, 66, 153
  relation to infinite divisibility, 153, 154
state-space
  formalism, 138
  model, 95
stationarity, 3, 31
  non-, 36
  strict sense, 45
  wide sense, 5, 45
  of SDE, 64
step number, 40, 165
  fluctuations in, 40, 43, 52, 53, 68, 71, 88, 166
stochastic differential equation (SDE), 1, 5, 13, 20–23, 25–27, 31–35, 40–44, 49, 52–54, 56–59, 62–64, 67, 69, 71, 72, 74, 86, 91–93, 97, 99, 100, 105–109, 111, 113, 117–119, 121, 123, 124, 126, 129–132, 137–139, 141–145, 147–151, 156, 157, 159, 165–168
stochastic differential equation SDE, 139
stochastic equilibrium, 18, 27
Stratonovich integral, 8, 155
  chain rule for, 156
superposition, 68, 70, 90, 144, 153
  coherent, 25
  principle of, 145
  weak, 91
Swerling
  target in Rayleigh clutter, 71

target returns, 97, 114, 115
time reversal symmetry, 18, 19
Tough
  Robert, 99

vector scattering, 12, 27, 30, 36–38, 72, 73, 76, 86
  correlation in, 35
volatility, 7, 8
  instantaneous observability of, 27, 28, 33, 34, 124, 134, 159
  tensor transformation of, 15

Weibull distribution, 29, 67
Wiener process, 7, 9, 157, 159
wireless propagation, 88, 95, 141
  SDE models for, 95, 141

z-parameter, 138, 140, 141, 147–149